NUMERICAL METHODS IN ENGINEERING AND SCIENCE

NUMERICAL METHODS IN ENGINEERING AND SCIENCE

Carl E. Pearson

University of Washington

 VAN NOSTRAND REINHOLD COMPANY

————————————————————————— New York

Manufactured in the United States of America.

Published by Van Nostrand Reinhold Company Inc.
115 Fifth Avenue
New York, New York 10003

Van Nostrand Reinhold Company Limited
Molly Millars Lane
Wokingham, Berkshire RG11 2PY, England

Van Nostrand Reinhold
480 Latrobe Street
Melbourne, Victoria 3000, Australia

Macmillan of Canada
Division of Gage Publishing Limited
164 Commander Boulevard
Agincourt, Ontario MIS 3C7, Canada

15 14 13 12 11 10 9 8 7 6 5 4 3 2 1

Library of Congress Cataloging-in-Publication Data
Pearson, Carl E.
   Numerical methods in engineering and science.
   Includes bibliographies and index.
   1. Engineering mathematics.   2. Science—Mathematics.
3. Numerical analysis.   I. Title.
TA335.P43   1986   620'.0042   85-22516
ISBN 0-442-27344-4

Contents

Preface

A course in numerical analysis has become accepted as an important ingredient in the undergraduate education of engineers and scientists. *Numerical Methods in Engineering and Science* reflects my experience in teaching such a course for several years. Related work in industry and research has influenced my choice of content and method of presentation.

Most students at the undergraduate level will have had, at the very least, an introductory course in ordinary differential equations. Tutorial appendixes on complex variables, determinants, partial differentiation, and Taylor expansions are included at the end of this book. Other background material is developed as needed. For example, Chapter 2 (linear equations) begins with an outline of matrix algebra; this will represent review material for some students, but it will spare others the necessity of consulting references. Proofs for almost all results of importance are given, in what I hope is palatable form. Overall, the book should be reasonably self-contained.

A number of illustrative computer programs are provided. The language chosen is FORTRAN (ANSI 77) because of the wide availability of service and application programs in that language. I am aware of course that some readers, for good reasons, will prefer other languages; however, since most languages are sufficiently similar to FORTRAN there should be little difficulty in translating from one language to another as required.

I have not tried to include an extensive collection of library-type programs. At this stage, it seems to me that it is important for the student to acquire facility in writing actual programs (and this is called for, in the text and in exercises). This kind of programming experience should help solidify the understanding of numerical techniques, as well as provide perspective for the eventual use of subroutine libraries.

This book contains more material than would normally be included in an introductory undergraduate course. I feel, however, that it is useful to cover the various topics with some degree of completeness so that the book may serve the student as a subsequent reference and source of ideas. Illustrative examples are given throughout the text. It is important for the student to work problems, and a fairly large number of problems that illustrate (and in some cases, extend) the material of the text will be found at the end of each chapter. In a course based on this book, the instructor may want to supplement these problems with some of the usual kind of drill-type exercises.

I am grateful to many past and present associates who have had an influence on *Numerical Methods in Engineering and Science*. Several colleagues have been kind enough to critically read the manuscript, in various stages of preparation, and to offer suggestions concerning appropriate material or treatment. I want to express particular appreciation to Professors David Benney (Massachusetts Institute of Technology), Graham Carey (University of Texas), Walter Christiansen (University of Washington), Robin Esch (Boston University), Robert MacCormack (Stanford University), and Chris Newbery (University of Kentucky). The responsibility for inaccuracies or for inelegances of exposition remains, of course, mine alone.

It is a pleasure to thank Kathy Hamilton for her painstaking efforts to make everything legible and for her patience in dealing with many changes. In the process, she has become something of a numerical analyst herself.

CARL E. PEARSON

1

NONLINEAR EQUATIONS

The purpose of this chapter is to discuss a number of methods applicable to the solution of a single nonlinear equation, usually algebraic or transcendental in character. Sets of equations are considered in Chapter 2 (linear case) and Chapter 3 (nonlinear case).

1.1 A SAMPLE PROBLEM

Suppose that electrical cables are to be strung between a series of towers. To design the towers one has to know the tension in the cables, and this depends on the ground clearance desired. Consider such a cable, as shown in Figure 1.1. It is a standard exercise in differential equations texts to show that the cable takes the shape of a catenary. If ρ denotes the linear density of the cable, T the horizontal component of the cable tension, and g the acceleration of gravity, then in terms of horizontal distance s, the height w of the cable is given by

$$w = \frac{1}{\alpha} [\cosh \alpha s - 1]$$

$$= \frac{1}{\alpha} \left[\frac{e^{\alpha s} + e^{-\alpha s}}{2} - 1 \right], \tag{1.1}$$

where $\alpha = \rho g / T$. Here the origin of the (s, w) coordinate system has been made to coincide with the lowest point on the cable, as shown in Figure 1.1.

If l denotes the half-span, then at $s = l$ the height of the cable above its lowest point is given by

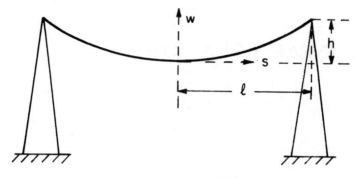

Figure 1.1. Cable problem.

$$h = \frac{1}{\alpha} [\cosh \alpha l - 1] .$$ (1.2)

The problem is then to determine α (and so T) if the sag h is specified. It is often useful to introduce dimensionless variables, and we do so here by the definitions

$$x = \alpha l = \frac{\rho g l}{T} , \qquad \lambda = \frac{h}{l} .$$

The quantities x and λ are dimensionless. Equation (1.2) becomes

$$\lambda x = \cosh x - 1 .$$ (1.3)

This equation is now to be solved for x, where λ is specified. This problem will serve as an example to which several solution methods will be applied. The problem is, of course, a rather simple one, but it has the advantage that we can concentrate on solution methods without encumbrances of algebraic complexity.

It is worthwhile to begin with a graphical look at Equation (1.3). Suppose, to be specific, that the designer's choice of cable clearance transforms into a desired value of .1575 for λ. Then we could plot the two curves

$$y = 0.1575x , \qquad y = \cosh x - 1$$

and find that value of x at which the curves intersect. This is done in Figure 1.2 and we obtain $x \cong .32$. A refinement of the graph would give a more accurate value for x, but this process could become cumbersome if several significant figures were required, if the equation to be solved were a complicated one, or if solutions corresponding to several values of λ were required. Consequently, it is appropriate to look for more efficient methods. Nevertheless, a simple preliminary sketch is often useful—it can protect us against a future gross error and it can also provide a starting value for an iterative process.

We remark that it is sometimes useful to interchange the roles of dependent and independent variables. Our interest is in solving Equation (1.3) for x if λ is specified. We could equally well think of Equation (1.3) as determining λ when x

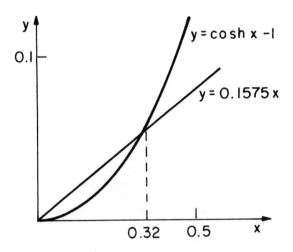

Figure 1.2. Graphical determination of x.

is given. It would be easy to plot λ as a function of x; if a number of values of λ are to be considered, the corresponding values of x could be read off from this graph. Of course, this particular equation happens to be of rather simple form— the parameter λ could easily appear on both sides of a more complicated equation.

1.2 REPEATED BISECTION METHOD

Define

$$y = \cosh x - \lambda x - 1 . \tag{1.4}$$

Then the problem of Section 1.1 requires us to find that value of x, say x_0, for which y vanishes. Again, we take $\lambda = .1575$.

The idea of the bisection method is to start with a pair of values for x, say x_1 and x_2, for which the corresponding values for y (denoted by y_1 and y_2, respectively) are of opposite sign. Then x_0 must lie between x_1 and x_2. We calculate next the midpoint value $x_3 = \frac{1}{2}(x_1 + x_2)$ and determine the corresponding quantity y_3. If y_3 has the same sign as y_1, we deduce that x_0 must lie between the pair x_2 and x_3, whereas if y_3 has the same sign as y_2, then x_0 must lie between x_1 and x_3. Of these two subintervals the one that is known to contain x_0 is then bisected again, and the process continues iteratively.

Figure 1.3 shows the first few steps for the case of Equation (1.4), with $\lambda = .1575$. Guided by the approximate value .32 for x_0, we chose x_1 to be .31 and x_2 to be .33. Calculation shows that $y_1 < 0$ and $y_2 > 0$ (the exact values don't matter much—only the signs), so that x_0 must lie between x_1 and x_2. The midpoint value x_3 is given by $\frac{1}{2}(x_1 + x_2) = .32$, and we find $y_3 > 0$, so that x_0 must lie between x_1 and x_3. The midpoint of these two is $x_4 = \frac{1}{2}(x_1 + x_3) = .315$, for

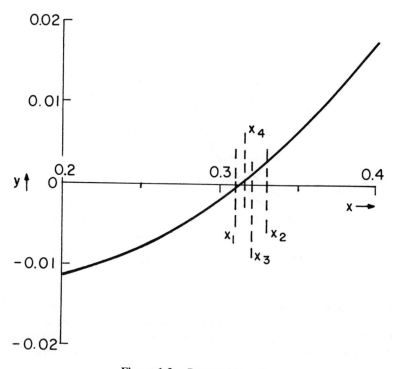

Figure 1.3.   Repeated bisection.

which $y_4 > 0$. Then $x_5 = .3125$ (with $y_5 > 0$, so that x_0 must lie between .31 and .3125), and we can continue indefinitely, halving the size of the interval containing x_0 at each step.

Proceeding as far as x_8, we find that x_0 must lie somewhere in the interval $(.3124, .3125)$. More generally, if the original interval length is denoted by δ (in our case, $\delta = x_2 - x_1 = .02$), then after n bisections the interval containing the solution point will have length $\delta/2^n$. Although we will shortly look at more efficient methods, the bisection method has some advantages. At each step, only functional evaluations (and in fact only one new one) are necessary; we do not have to calculate derivatives. Also, convergence is guaranteed, since we have a sequence of intervals, of decreasing size, within which x_0 must lie.

If y is a more intricate function than that described by Equation (1.4), the initial interval may contain several zeros of that function. In that event only one of those zeros will usually be found by the bisection method.

1.3 SECANT METHODS

Let $y = f(x)$, where $f(x)$ is some given function, and let it be required to find that value of x, say x_0, at which y vanishes. A plot of $y = f(x)$ might look something like that shown in Figure 1.4. In the secant method one chooses two x-values, x_1 and x_2, and calculates the corresponding y_1 and y_2 values. This gives a pair of points (x_1, y_1) and (x_2, y_2) lying on the curve. The line joining these points (the *secant*) is drawn, and its intersection point x_3 with the x-axis is calculated. Figure 1.4 suggests that if x_1 and x_2 are reasonably close to the desired root x_0, then x_3 should be an even better approximation to x_0. We can now iterate the process, using the pair (x_2, y_2), (x_3, y_3) for the next step, and so on.

In the case of Equation (1.4), with $\lambda = .1575$ as before, let us take $x_1 = .34$, $x_2 = .33$. Then the straight line through $(.33, .002971)$ and $(.34, .004809)$ is given by

$$y = .002971 + \left(\frac{.004809 - .002971}{.34 - .33}\right)(x - .33) \, .$$

The point x_3 at which this line cuts the x-axis is that value of x for which $y = 0$; we find $x_3 = .3138$. One more step (starting with x_2 and x_3) yields $x_4 = .3125$. Further steps change this value only slightly, so that we can take x_4 as an approximation to x_0, correct to about four figures.

The above method is termed the *secant method*, and the example shows that it can be very effective. Unfortunately, pathological situations, such as the presence of extrema (see Fig. 1.5), can arise in which the method may not converge.

A related method, with guaranteed convergence, is the *rule of false position—* or *regula falsi*, as it was termed by seventeenth-century numerical analysts. In this

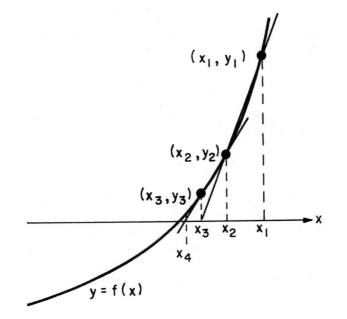

Figure 1.4. Secant method.

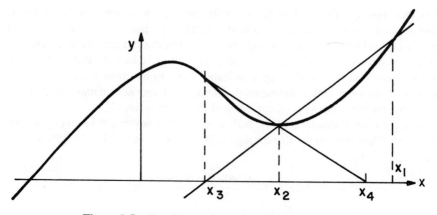

Figure 1.5. Possible nonconvergence of secant method.

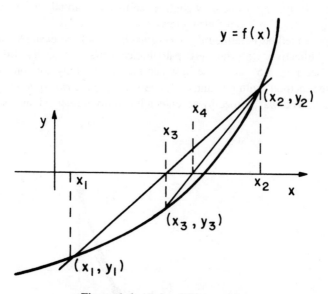

Figure 1.6. Rule of false position.

method the two initial x-values, x_1 and x_2, are chosen to lie on opposite sides of the root x_0—that is, y_1 and y_2 must have opposite signs. The two points (x_1, y_1) and (x_2, y_2) are again connected by a straight line, which intersects the x-axis at x_3. The next iteration step starts with that one of the two possible pairs—the pair (x_1, y_1) and (x_3, y_3), or the pair (x_2, y_2) and (x_3, y_3)—for which the two y-values have opposite signs. The method now continues in the same way. (See Fig. 1.6). If we take $x_1 = .31$, $x_2 = .33$ in our standard example, we find $x_3 = .3123$, $x_4 = .3124$, $x_5 = .3124$, ...

As indicated in Figure 1.6, the successive iteration points x_i will eventually all lie on the same side of the root x_0. Variants of the method, which provide sets of x_i-values approaching x_0 from both above and below, are possible; see Problem 1.7.

The subroutine FALSE is based on the rule of false position. The input and output parameters are described in comment statements. (Note that the desired tolerance, TOL, should not be chosen so small that the number of significant figures carried by the computer becomes crucial.) A second "driver" program follows the subroutine. This program uses FALSE to again solve for x in $\cosh(x) - .1575x - 1 = 0$. The output is found to be

$$.31244E+00 \qquad -.11921E-0.5 \qquad 2$$

which checks the previous answer and also the subroutine program itself.

```
C   SUBROUTINE FALSE(N,XL,XR,XC,FC,I,TOL) USES METHOD OF FALSE
C   POSITION TO ITERATE TOWARDS A ZERO OF A FUNCTION FF(X).  THE
C   ROUTINE WILL ITERATE N TIMES UNLESS FF(X) BECOMES LESS THAN
C   TOL IN ABSOLUTE VALUE.  INPUT VALUES OF X FOR WHICH FF(X) HAS
C   OPPOSITE SIGNS ARE REQUIRED.
C
C   INPUT:
C         XL,XR = VALUES OF X FOR WHICH FF(X) HAS OPPOSITE SIGNS,
C             AND BETWEEN WHICH A ZERO OF FF(X) MUST LIE
C             N = MAXIMUM NUMBER OF ITERATIONS
C           TOL = TOLERANCE.  ROUTINE ITERATES N TIMES, UNLESS
C             ABSOLUTE VALUE OF FF BECOMES LESS THAN TOL
C
C   OUTPUT:
C            XC = LAST ITERATION POINT, AND BEST APPROXIMATION
C                 TO VALUE OF X SUCH THAT FF(X)=0
C            FC = VALUE OF FF(XC)
C            I = NUMBER OF ITERATIONS ACTUALLY PERFORMED
C
C   FUNCTION CALLED:  FF(X)
C
        SUBROUTINE FALSE(N,XL,XR,XC,FC,I,TOL)
        I=0
        FL=FF(XL)
        FR=FF(XR)
3       I=I+1
        XC=(XL*FR-XR*FL)/(FR-FL)
        FC=FF(XC)
        IF(ABS(FC).LE.TOL.OR.I.GE.N) RETURN
C   FOR NEXT ITERATION, CHOOSE TWO POINTS BRACKETING ZERO OF FF
        IF(FL*FC.LT.0.) THEN
        XR=XC
        FR=FC
        ELSE
        XL=XC
        FL=FC
        END IF
        GO TO 3
        END
```

```
C   THE FOLLOWING PROGRAM USES <FALSE> TO APPROXIMATE A ZERO
C   OF THE FUNCTION      COSH(X)-.1575*X-1
C
    PROGRAM TEST
    N=5
    XL=.31
    XR=.33
    TOL=1.E-5
    CALL FALSE(N,XL,XR,XC,FC,I,TOL)
    WRITE(*,100) XC,FC,I
100 FORMAT(2E12.5,I4)
    END
    FUNCTION FF(X)
    FF=COSH(X)-.1575*X-1.
    END
```

1.4 NEWTON'S METHOD

In the secant method of Figure 1.4, a line was drawn through two points (x_1, y_1) and (x_2, y_2) of a curve $y = f(x)$, and the intersection of this line with the x-axis was determined. If the point x_2 is made to approach x_1, then in the limit the secant becomes the tangent to the curve at the point (x_1, y_1), and this leads to *Newton's method* (sometimes called the Newton–Raphson method).

We use a prime to denote a derivative: so if $y = f(x)$, then $dy/dx = f'(x)$. Then the slope of the tangent line at x_1 is $f'(x_1)$, and the equation of this line is

$$y = f(x_1) + f'(x_1)(x - x_1) .$$

Let x_2 denote the intersection point of this line with the x-axis (see Fig. 1.7). Then

$$x_2 = x_1 - \frac{f(x_1)}{f'(x_1)} . \tag{1.5}$$

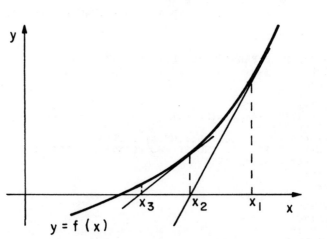

Figure 1.7. Newton's method.

The process can now be repeated, with

$$x_3 = x_2 - \frac{f(x_2)}{f'(x_2)} ,$$

and so on. In general,

$$x_{i+1} = x_i - \frac{f(x_i)}{f'(x_i)} . \tag{1.6}$$

For the previous example, with $f(x) = \cosh(x) - .1575x - 1$, Eq. (1.6) gives

$$x_{i+1} = x_i - \frac{\cosh x_i - .1575x_i - 1}{\sinh x_i - .1575} .$$

Starting at $x_1 = .30$, we obtain the sequence $x_2 = .31300$, $x_3 = .31245$, and at this point the sequence has essentially converged.

It is easy to derive Equation (1.6) analytically by considering the change in $f(x)$ resulting from a small change in x; also, this approach will carry over to the several-variable case of Chapter 3. Suppose x_i has been determined, and let x be any point near x_i. From the definition of a derivative, we know that if we define $\epsilon_i(x)$ by

$$\epsilon_i(x) = f'(x_i) - \frac{f(x) - f(x_i)}{x - x_i} . \tag{1.7}$$

then $\epsilon_i(x) \rightarrow 0$ as $x \rightarrow x_i$. This equation may be written

$$f(x) - f(x_i) = f'(x_i) \cdot (x - x_i) - \epsilon_i(x) \cdot (x - x_i) . \tag{1.8}$$

Since $\epsilon_i(x) \rightarrow 0$ as $x \rightarrow x_i$, it follows that a first approximation to $f(x) - f(x_i)$, if x is close to x_i, is given by the term $f'(x_i) \cdot (x - x_i)$. Within this approximation, we choose x, denoted now by x_{i+1}, so as to make $f(x)$ vanish; Equation (1.8) becomes

$$0 - f(x_i) \cong f'(x_i) \cdot (x_{i+1} - x_i) ,$$

and this leads again to Equation (1.6).

Newton's method converges very rapidly, once the iteration points are close enough to the root. To show this, we use Taylor's theorem, the derivation of which is sketched in Appendix A. Suppose x_0 is a root of $f(x) = 0$, and suppose the ith iteration point, x_i, is close enough to x_0 that the higher-order terms in Taylor's formula

$$f(x_i) = f(x_0) + f'(x_0) \cdot (x_i - x_0) + \tfrac{1}{2} f''(x_0) \cdot (x_i - x_0)^2 + \text{higher-order terms} \tag{1.9}$$

are negligible. Then, since $f(x_0) = 0$, Equation (1.6) becomes

$$x_{i+1} \cong x_i - \frac{f'(x_0) \cdot (x_i - x_0) + \tfrac{1}{2} f''(x_0) \cdot (x_i - x_0)^2}{f'(x_0) + f''(x_0) \cdot (x_i - x_0)} ,$$

where the denominator represents the corresponding Taylor expansion for $f'(x_i)$. This equation can be rewritten as

$$x_{i+1} \cong x_i - \frac{(x_i - x_0) + \frac{1}{2}\beta(x_i - x_0)^2}{1 + \beta(x_i - x_0)},$$

where β denotes $f''(x_0)/f'(x_0)$; we assume here that $f'(x_0) \neq 0$. Finally, for small values of $x_i - x_0$, the approximation

$$\frac{1}{1 + \beta(x_i - x_0)} \cong 1 - \beta(x_i - x_0)$$

can be used to give

$$x_{i+1} \cong x_i - [(x_i - x_0) + \tfrac{1}{2}\beta(x_i - x_0)^2] \cdot [1 - \beta(x_i - x_0)]$$

$$\cong x_i - (x_i - x_0) + \tfrac{1}{2}\beta(x_i - x_0)^2$$

within second-order terms. This equation may be rewritten in the form

$$x_{i+1} - x_0 \cong \frac{1}{2}\frac{f''(x_0)}{f'(x_0)}(x_i - x_0)^2 \,. \qquad \textbf{(1.10)}$$

Since $x_i - x_0$ represents the error at the ith iteration and $x_{i+1} - x_0$ that at the $(i + 1)$st iteration, it follows that the error at the $(i + 1)$st step is proportional to the square of the error at the ith step. We can expect, roughly, to double the number of significant figures of accuracy at each new iteration step. Equation (1.10) was obtained on the assumptions that $f'(x_0)$ is nonzero and that x_i is close enough to x_0 that the above approximations are valid.

This "quadratic" convergence rate frequently makes Newton's method the method of choice for nonlinear equations (cf. Prob. 1.9 and 1.10, which concern the convergence rates for other methods). However, one pays a price—at each step, both $f(x_i)$ and $f'(x_i)$ must be evaluated, and if $f(x)$ is a complicated function, $f'(x_i)$ could be rather unwieldly. A simple alternative to the evaluation of $f'(x_i)$ is to choose a point $(x_i + \epsilon)$ very close to x_i and to use

$$f'(x_i) \cong \frac{f(x_i + \epsilon) - f(x_i)}{\epsilon} \,. \qquad \textbf{(1.11)}$$

Newton's method works perfectly well for complex-valued roots. Suppose that $F(z)$ is a complex-valued function of the complex variable z. Complex numbers are reviewed in Appendix C; however, all we need at the moment is the fact that complex numbers obey all the usual rules of algebra, and that the derivative of $F(z)$ is defined in exactly the same way as in the real-variable case, so that a result analogous to that of Equation (1.7) holds. Consequently, we obtain Equation (1.6) as before, except that complex numbers are involved:

$$z_{i+1} = z_i - \frac{F(z_i)}{F'(z_i)} \,. \qquad \textbf{(1.12)}$$

As a simple example, let $F(z) = z^2 - 2z + 5$, which has zeros at $1 \pm 2i$, where $i = \sqrt{-1}$. Then Equation (1.12) becomes

$$z_{i+1} = z_i - \frac{z_i^2 - 2z_i + 5}{2z_i - 2} . \qquad (1.13)$$

Take $z_0 = 1 - 2i$. The reader should verify that if z_1 is chosen anywhere near z_0, then the sequence $z_2, z_3, \ldots$, obtained from Equation (1.13) will converge very rapidly to z_0.

1.5 SUCCESSIVE SUBSTITUTION

For $\lambda = .1575$ the problem associated with Equation (1.4) may be rephrased as follows: Find a value of x such that

$$x = g(x) = \frac{\cosh x - 1}{.1575} . \qquad (1.14)$$

A natural iterative approach is to start with some tentative value x_1 and generate a sequence $x_2, x_3, \ldots$ by

$$x_{n+1} = g(x_n) . \qquad (1.15)$$

We can hope that this will converge to the desired root x_0. In the present case, the idea does not work; with $x_1 = .32$, we obtain $x_2 = .3279$, $x_3 = .3443$, $x_4 = .3801$, $x_5 = .4642$, and the sequence does not converge.

To see what went wrong, consider the general case of Equation (1.15), where $g(x)$ is some specified function. The desired root x_0 corresponds to the intersection point of the two curves $y = x$ and $y = g(x)$. The sequence of iteration points is easily obtained graphically; Figure 1.8a is for a case in which we get convergence,

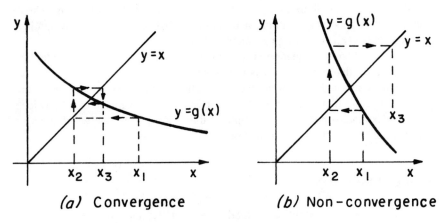

Figure 1.8. Iteration process for $x = g(x)$.

and Figure 1.8*b* is for a case in which we do not. It is clear from this graph that the condition for convergence is that, in the iteration region, $|g'(x)|$ must be bounded by some quantity $L < 1$.

We can also prove analytically that the process will converge for $L < 1$. Let x_0 denote the root. Then from Equation (1.15), we have

$$|x_{n+1} - x_0| = |g(x_n) - x_0| = |g(x_n) - g(x_0)| ,$$

since $g(x_0) = x_0$. But if $|g'(x)| < L < 1$ in the region of interest, then $|g(x_n) - g(x_0)| < L \cdot |x_n - x_0|$ (since the function $g(x)$ cannot change more rapidly, as we move away from x_0, than at the rate L). Thus

$$|x_{n+1} - x_0| < L \cdot |x_n - x_0| ,$$

so that the distance of each new iteration point from x_0 must be less than L times the preceding such distance. Since $L < 1$, we do have convergence. The reader may verify that the function $g(x)$ defined by Equation (1.14) fails to satisfy the criterion $|g'| < L < 1$.

Even if $g(x)$ does not satisfy the condition $|g'| < L < 1$, a modification in the iteration process may rescue it. Define

$$G(x) = Ax + (1 - A)g(x) , \tag{1.16}$$

where A is a constant to be determined, and consider the substitution process

$$x_{n+1} = G(x_n) .$$

Observe first that the equation $x = G(x)$ has the same solution x_0 as before, since Equation (1.16) gives $G(x_0) = Ax_0 + (1 - A)g(x_0) = g(x_0)$. However, we are now concerned with $|G'|$ rather than with $|g'|$; from Equation (1.16),

$$|G'(x)| = |A + (1 - A)g'(x)| .$$

Even if $|g'| > 1$ in a certain region, we can generally choose A so that $|A + (1 - A)g'| < L < 1$ (think of a plot of $A + (1 - A)g'$ versus A for some chosen value of g'). A good way to think of Equation (1.16) is that we are "weighting" the new values $g(x_n)$ with the previous values x_n.

For the problem of Equation (1.14), the reader should choose a suitable value for A and verify that convergence is now obtained.

In connection with the successive substitution method, we remark that it may be useful to reformulate the substitution process to take advantage of a knowledge of the approximate solution. An example of this, for a problem in which a zero of a polynomial is sought, is given in Problem 1.4.

1.6 ACCELERATION

The successive substitution method of Section 1.5 encompasses more situations than might at first be apparent. For example, Newton's method can be put in the form

$$x_{n+1} = g(x_n) \tag{1.17}$$

if we simply define $g(x_n)$ [cf. Eq. (1.6)] by

$$g(x_n) = x_n - \frac{f(x_n)}{f'(x_n)}.$$

Similarly, one endpoint in the false position method normally becomes fixed, so that it also can be described in the form of Equation (1.17).

Consequently, Equation (1.17) occurs very commonly, and any method that can enhance the convergence process is of interest. One method—that of weighting—has already been discussed in connection with Equation (1.16); we now consider another method.

During the successive substitution process, we obtain a sequence of iterates $\ldots, x_n, x_{n+1}, x_{n+2}, \ldots$, and it is natural to ask whether we can use this sequence to "predict" the solution x_0. Since x_0 satisfies the equation $x_0 = g(x_0)$, we can subtract x_0 from both sides of Equation (1.17) to obtain

$$x_{n+1} - x_0 = g(x_n) - g(x_0).$$

If x_n is sufficiently close to x_0 so that the first term of a Taylor expansion is adequate, this equation can be rewritten as

$$x_{n+1} - x_0 \cong g'(x_0) \cdot (x_n - x_0).$$

Similarly,

$$x_{n+2} - x_0 \cong g'(x_0) \cdot (x_{n+1} - x_0).$$

The quantities $g'(x_0)$ and x_0 are unknown but can be obtained by solving these two equations (since x_n, x_{n+1}, x_{n+2} are known from the successive substitution process). In particular, we obtain

$$x_0 \cong \frac{x_n x_{n+2} - x_{n+1}^2}{x_n - 2x_{n+1} + x_{n+2}}. \tag{1.18}$$

This process is sometimes called *Aitken acceleration*. If the original successive substitution process is converging rather slowly, Equation (1.18) may provide a jump ahead to a better iteration value.

For a simple example, consider the process $x_{n+1} = .06/\sin x_n$, which converges slowly. Starting with $x_1 = .3$, we obtain, for example, $x_5 = .2952451$, $x_6 = .2062037$, and $x_7 = .2930467$, $x_8 = .2077056$. Convergence seems slow, so we use Equation (1.18), with $n = 6$, say, to obtain $x_0 \cong .250004$. Starting with this new point, taking two more iterates, and again using Equation (1.18), we obtain .24622, which is close to the solution value $x_0 = .24619$.

As a matter of perspective, we note that we could probably also have improved the convergence in this problem by averaging the iterates in pairs, and that use of Newton's method to start with would have been much more effective. However, alternatives such as these may not be readily available in more complicated problems, and acceleration via Equation (1.18) can then provide a useful tool.

1.7 DIFFICULTIES AND EXTENSIONS

We collect in this section a number of pertinent remarks concerning zero-finding methods:

1. Suppose there are a number of x-values for which $f(x)$ vanishes. Each of the methods discussed so far can be effective in locating a particular zero, provided we start the iteration process at a suitable point (i.e., close enough to that zero). We could presumably find all the other zeros by repeating the process, using, for example, graphical methods to suggest initial guesses. Sometimes, however, the zeros of $f(x)$ are tightly clustered, and even if we start at a point that is close to a second zero, we may find ourselves converging again to the first zero. One way around this, after a zero, say x_0, has been found is to define

$$g(x) = \frac{f(x)}{x - x_0}, \tag{1.19}$$

which (normally) would no longer vanish at x_0 but would still vanish at the remaining zeros. Then the method could proceed, with $f(x)$ replaced by $g(x)$. Once a second zero has been found, $g(x)$ could be replaced by a modified function obtained in the same way, and so on.

2. Our second remark concerns a difficulty that could arise in Newton's method. Suppose the function $f(x)$ is tangent to the x-axis at the point x_0, as in Figure 1.9. Then as $x_i \to x_0$, $f'(x_i) \to 0$, so that the correction to x_i given by Equation (1.6) can become very large. One way in which to evade this difficulty, if it seems to be arising, is to define

$$g(x) = \frac{f(x)}{f'(x)} \tag{1.20}$$

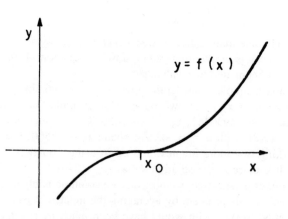

Figure 1.9. Tangency case.

and to apply Newton's method to the modified function $g(x)$. Near x_0 we have (from a Taylor series, using $f(x_0) = 0$, $f'(x_0) = 0$)

$$g(x) \cong \frac{\frac{1}{2} f''(x_0) \cdot (x - x_0)^2}{f''(x_0) \cdot (x - x_0)} \cong \frac{1}{2}(x - x_0) \, ,$$

so that $g(x)$ vanishes at $x = x_0$ and, moreover, is not tangent to the x-axis at x_0; thus, Newton's method should work well. We have assumed here that $f''(x_0) \neq 0$; if $f''(x_0)$ also vanishes, but $f'''(x_0)$ does not, we would have (again using a Taylor expansion)

$$g(x) \cong \frac{\frac{1}{6} f'''(x_0) \cdot (x - x_0)^3}{\frac{1}{2} f'''(x_0) \cdot (x - x_0)^2} \cong \frac{1}{3}(x - x_0) \, ,$$

and Newton's method should still work. The general idea is clear.

3. Continuing with Newton's method, we observe that even if $f'(x_0)$ is nonzero, it can happen that $f'(x_i)$ is very small, so that the correction term subtracted from x_i in Equation (1.6) can be uncomfortably large and so move us to a point x_{i+1} far from x_0. Figure 1.10 illustrates the situation. To obtain partial protection against this possibility, it is conventional to incorporate in a Newton's method computer algorithm a limit δ that $|x_{i+1} - x_i|$ is not permitted to exceed. Then if we find that $|f(x_i)/f'(x_i)| > \delta$, we obtain x_{i+1} from

$$x_{i+1} = x_i - \delta \left\{ \operatorname{sgn} \left[\frac{f(x_i)}{f'(x_i)} \right] \right\} , \qquad (1.21)$$

where the function $\operatorname{sgn}(w)$ is $+1$ if $w > 0$, and -1 if $w < 0$. A similar precaution is, of course, applicable to the secant method.

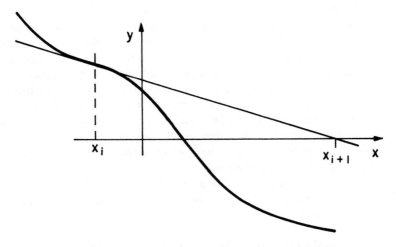

Figure 1.10. Effect of small slope in Newton's method.

4. A natural extension of the secant method, which approximates the true function $f(x)$ locally by means of a straight line, would be to use a parabola. Let $(x_1, f(x_1))$, $(x_2, f(x_2))$, $(x_3, f(x_3))$ be three points on the curve near a root x_0. Then the coefficients A, B, C of a parabola given by $y = A + Bx + Cx^2$ can be determined so that the parabola also passes through these three points (the three equations determining A, B, C are typified by that for the first point, which is $f(x_1) = A + Bx_1 + Cx_1^2$), and then we can find the intersection point of the parabola with the x-axis by the usual formula for the root of $A + Bx + Cx^2 = 0$. This is *Muller's method*. It can provide more rapid convergence than the secant method, but it has the disadvantages that (a) one has to decide which of the two roots of the quadratic equation is appropriate—usually, of course, the closest one—and (b) in some cases the roots of the quadratic could be complex, and that would not be very helpful. One could also reverse variables and write x as a parabolic function of y or use a conic-section curve or a higher-order polynomial. (*Warning:* High-order polynomials have a tendency to exhibit oscillations.)

5. The efficiency of any iterative root-finding algorithm depends not only on the convergence speed of that algorithm but also on the accuracy of the starting guess. Consequently, it can be worthwhile to invest some effort in obtaining a good initial guess. As an example, suppose that we were designing a square root algorithm for a computer's library of intrinsic functions. Newton's method is again suitable; given some positive number N, we want to find x such that $f(x) = x^2 - N = 0$. Equation (1.6) gives

$$x_{i+1} = x_i - \frac{x_i^2 - N}{2x_i} = \frac{1}{2}\left(x_i + \frac{N}{x_i}\right). \tag{1.22}$$

(Eq. (1.22) is intuitively plausible, because if, for example, $x_i < \sqrt{N}$, then N/x_i will be $> \sqrt{N}$, so the average of these two quantities should be an improvement.) Since this algorithm can be expected to be used very many times, we want to make it efficient, and so we want a good value for x_1. If preliminary scaling or normalization has been done so that $1 < N < 2$, say, then we could plot $\sqrt{N}$ against N and try to fit some simple function (or functions) to this curve. For example, we could use one linear function of N for the first half of the range, and another for the second half in order to provide a good initial choice x_1. We could experiment with different choices of this kind and determine the corresponding number of iterations required for convergence. We would try to find an initial guess x_1 such that a small fixed number of iterations would guarantee adequate convergence without the time-wasting necessity of having to perform a test at each step to see if the current value of x_i is good enough.

6. Until now, we have implicitly assumed that a computer carries out arithmetical operations exactly. Because of the finite word length of a computer, this is of course not correct, and there are situations in which round-off error must be considered.

A computer carrying about 7 significant figures was used to add together 1,000 numbers, each equal to .001. The result was found to be .9999907. Using double precision (14 significant figures) on the same computer, it was similarly found that when 100,000 numbers, each equal to 1×10^{-5}, were added, the error in the result was about 1×10^{-12}. More generally, it is sometimes remarked that the axioms of arithmetic are not applicable to computer operations. Thus, if x, y, z are real numbers to be added together, the result may depend on the order of operations; we need not have $(x + y) + z = x + (y + z)$.

The effects of finite word length are usually important only when many operations are to be carried out. However, there are situations—certain matrix operations, or differential equation algorithms, for example—where round-off error can have unexpectedly large effects, even if the total number of operations is small. We will consider those cases as they arise.

Sometimes, a reformulation of the problem can minimize error. For example, the evaluation of $f(x) = 1 - \cos x$, for small values of x, leads to the subtraction of one number from another of almost equal value, with a consequent loss in significant figures. The use of the equivalent formula $f(x) = 2 \sin^2(x/2)$ avoids this difficulty.

ANNOTATED BIBLIOGRAPHY

B. Carnahan, H. A. Luther, and J. O. Wilkes, 1969, *Applied Numerical Methods*, John Wiley, New York, 604p.

   Chapter 3 deals with nonlinear equations, including an extensive survey of polynomial equations, and gives a large number of examples, with programs.

G. E. Forsythe, M. A. Malcolm, and C. B. Moler, 1977, *Computer Methods for Mathematical Computations*, Prentice-Hall, Englewood Cliffs, New Jersey, 259p.

   Although this text does have a chapter on nonlinear equations, the main reason for including it here is its very useful discussion of computer arithmetic—effects of word length, roundoff, sensitivity, and so forth. Some of the examples are provocative, some are amusing, and all are instructive. Chapters 1 and 2 are easy to read and well worthwhile.

H. H. Goldstine, 1977, *A History of Numerical Analysis*, Springer-Verlag, New York, 348p.

   The history of the Newton–Raphson procedure (p. 64ff) is of particular interest. Newton first used the method in 1669 in connection with the cubic equation $x^3 - 2x - 5 = 0$. Raphson gave a more general discussion in 1690 but acknowledged Newton's precedence.

A. S. Householder, 1970, *The Numerical Treatment of a Single Nonlinear Equation*, McGraw-Hill, New York, 216p.

   Look at this scholarly text if you want to see how sophisticated the subject can become. It is primarily a book for the specialist; however, it does provide an extensive bibliography.

PROBLEMS

1.1 Write a subroutine that uses the bisection method to solve $f(x) = 0$. Assume that the program is entered with values of x, denoted by XL and XR, for which $f(x)$ has opposite signs. Include a tolerance limit for $f(x)$ (as in the program of Sec. 1.3), and run a test case for which the solution is known.

1.2 In determining natural frequencies of vibration in mechanical or electrical systems, it is often necessary to solve transcendental equations of the kind exemplified by Equation (1.3). As one example of this kind find all real values of x, to five significant figures, satisfying $\tan x + 2x = 0, 0 \le x \le 2$. Use a hand calculator and your choice of any two of the methods discussed in this chapter.

1.3 The van der Waals equation of state for a gas is

$$\left(p + \frac{3}{v^2}\right)\left(v - \frac{1}{3}\right) = \frac{8}{3}T,$$

where p, v, and T are the "reduced" values of pressure, volume, and temperature, respectively (i.e., measured as ratios of actual physical values to their values at the critical point). Determine v if $p = 1.932$, $T = 1.311$. Use repeated bisection, starting with $v_1 = 1.2$, $v_2 = 1.5$. Next, try the iteration process, obtained by rewriting the above equation as

$$v_{i+1} = \frac{1}{3} + \frac{8T/3}{p + 3/v_i^2}$$

and explain your results. Note, incidentally, that the original equation could have been rewritten as a cubic in v (by multiplying through by v^2) had such a form been desired.

1.4 One iterative method for finding a root of the polynomial equation $x^4 - 3.07x^3 + x - 1.54 = 0$ proceeds as follows. After some experimentation we find that there seems to be a root near $x = 3$. Then make the change in variable $\xi = x - 3$, so that the equation becomes

$$(\xi + 3)^4 - 3.07(\xi + 3)^3 + (\xi + 3) - 1.54 = 0,$$

or

$$\xi^4 + 8.93\xi^3 + 26.37\xi^2 + 26.11\xi - .43 = 0.$$

Since ξ should be small, this now suggests the iterative process

$$\xi_{i+1} = \frac{.43}{26.11} - \frac{\xi_i^4 + 8.93\xi_i^3 + 26.37\xi_i^2}{26.11}.$$

Starting with $\xi_1 = 0$, find ξ_2, ξ_3, and ξ_4, and so an approximation to the desired root x of the original equation. This process could be described as "translating" a zero. Use a similar method to find the other real root of the original equation. Would this method work for a complex zero?

1.5 Write a computer subroutine implementing Newton's method. Incorporate the protection described in Section 1.7 against too large a step. Use it for the polynomial equation considered in Problem 1.4. Start with $x_1 = 3.5$. Also try the starting value $x_1 = 2.253$. Explain your results.

1.6 Explore Muller's method (Remark 4 of Sec. 1.7), comparing it with other techniques for some problems of known solution. Invent also a situation in which it could lead you astray. In general, how would you program an algorithm to choose the better of the two possible roots of the quadratic equation?

1.7 A modification in the method of false position, which leads to sets of points converging to the root of $f(x) = 0$ from both sides, could be obtained in several ways. One possibility is to define a second function $g(x) = f(x) + \alpha[f(x)]^2$, where the constant α is chosen so that the curvature g'' has opposite sign to f''; since g and f both vanish at x_0, the application of the method to both f and g would yield the desired two sequences. A second possibility is to alternate the method of false position with the bisection method. A third is to join a point such as $(x_3, f(x_3))$ in Figure 1.6, not to $(x_2, f(x_2))$, but rather to $(x_2, \frac{1}{2} f(x_2))$. Explore these ideas.

1.8 Use Newton's method to find the real and complex roots of

$$x^4 + 3.37x^3 + 8.566x^2 + 19.613x + 16.79 = 0.$$

Watch out for near-tangency.

1.9 In Section 1.4 a Taylor expansion was used to show that the convergence rate for Newton's method becomes quadratic as the root is approached—that is, each new error is proportional to the square of the preceding one. Use a similar method to show that the convergence rate for the false position method is generally linear—each new error is a certain constant times the preceding one. One can often simplify calculations of this kind by shifting axes so that the true zero is at $x = 0$.

1.10 Extend Problem 1.9 to show that for the secant method each new error becomes proportional to the product of the two preceding ones as the root is approached. Thus, the convergence rate is better than that for the false position method but not quite as good as in Newton's method.

1.11 Let $f(x) = \sqrt{x - x_0}\, g(x)$, where x_0 is a given number and $g(x)$ is a well-behaved function that is nonzero near x_0. How would various root-finding methods work in this case?

1.12 Does the secant method work for complex numbers? Explain. (*Hint:* See Prob. 1.10).

1.13 Rather than start with $f(x) = x^2 - N$ in devising the algorithm of Equation (1.22), we could have started with $f(x) = x - N/x$. Which approach requires fewer machine operations per iteration? (Note, incidentally, that division or multiplication by 2, with binary digits, requires only a register shift, so it is very fast). Devise a Newton's method algorithm for finding the *p*th root of any positive number *N*, where *p* is a positive integer.

1.14 The angle θ through which a supersonic airstream is turned by an oblique shock wave depends on the Mach number *M* of the incoming stream (ratio of stream velocity to speed of sound) and on the angle β that the shock wave makes with the original stream direction. The formula is

$$\tan \theta = 2(\cot \beta) \frac{M^2 \sin^2 \beta - 1}{M^2(1.4 + \cos 2\beta) + 2}.$$

If θ is measured as 7.13° for *M* = 1.43, use Newton's method to determine the two possible values of β. (This is a situation in which differentiation is cumbersome, so use an appropriately modified version of Newton's method.)

1.15 Suppose that $\alpha < g' < \beta$, where α and β are known. What is the optimal choice for *A* in the weighted iteration method of Equation (1.16)? One way in which to estimate g' is by use of a secant line; use this idea to construct an appropriate subroutine. Discuss precautions and convergence rate, and try some examples. Does the general method work for complex numbers? Explain.

1.16 The turbine temperature ratio *y* corresponding to minimum fuel consumption in a turboprop engine satisfies the equation $y = Ay^\epsilon + (B + \beta y)^2$, where the nondimensional engine parameters *A*, *B*, β, and ϵ have typical values .4, .3, .1, and .1, respectively. Assuming that the parameters lie within 25% of these typical numbers, devise a computer program that chooses an initial guess for *y* and obtains 5-digit accuracy after *n* iterations, where *n* (= ?) is fixed.

1.17 Figure 1.11 shows a planar mechanical linkage that transforms an input rotation ϕ about the axis *A* into an output rotation angle θ about the axis *B*. Write a subroutine to compute θ for given values of the lengths *a*, *b*, *c*, *d*, and for a given value of ϕ. Include a test to guard against inconsistent data. Plot a curve of θ versus ϕ for *a* = 5, *b* = 3, *c* = 2, *d* = 1. (There are two possible θ values for each value of ϕ; consider only the case $\gamma < 180°$.)

1.18 The free vibrations of a cantilever beam satisfy the equation $\cosh \alpha \cdot \cos \alpha = -1$, where the dimensionless parameter α is proportional to the square root of the natural frequency. Find the three smallest positive values of α satisfying this equation.

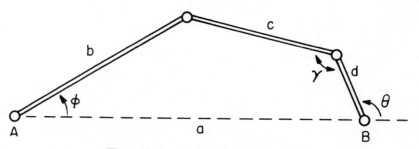

Figure 1.11. Linkage for Problem 1.17.

SIMULTANEOUS LINEAR EQUATIONS

2.1 MATRICES

A matrix is a rectangular array of real or complex numbers that may be added to or multiplied by other such arrays according to certain rules. We say that A is an *(m × n) matrix* if A has m rows and n columns; denoting the number, or *element*, in the ith row and jth column by a_{ij}, we have

$$A = \begin{bmatrix} a_{11} & a_{12} & \cdots & a_{1n} \\ a_{21} & a_{22} & \cdots & a_{2n} \\ \vdots & & & \vdots \\ a_{m1} & a_{m2} & \cdots & a_{mn} \end{bmatrix}. \tag{2.1}$$

The matrix A is said to be *square* (of order n) if $m = n$. If k is any *scalar* (i.e., any real or complex number), then kA is defined to be the new matrix obtained by multiplying each element in A by k. For example, with $i = \sqrt{-1}$,

$$3 \cdot \begin{bmatrix} 1 + i & -1 & 0 \\ i & 1 & -1 \\ 1 & 2 & 3 \end{bmatrix} = \begin{bmatrix} 3 + 3i & -3 & 0 \\ 3i & 3 & -3 \\ 3 & 6 & 9 \end{bmatrix}.$$

The sum of two $(m \times n)$ matrices, $A + B$, is defined to be the new $(m \times n)$ matrix C obtained by adding corresponding elements. If a_{ij}, b_{ij}, c_{ij} are the elements in the ith row and jth column of A, B, C, respectively, then $c_{ij} = a_{ij} + b_{ij}$. Thus

$$\begin{bmatrix} 0 & 1 \\ -1 & 2 \\ 3 & 0 \end{bmatrix} + \begin{bmatrix} -1 & -1 \\ 0 & 2 \\ -1 & 0 \end{bmatrix} = \begin{bmatrix} -1 & 0 \\ -1 & 4 \\ 2 & 0 \end{bmatrix}.$$

From these two definitions, it follows that multiplication by a scalar k is *distributive*, in the sense that if $C = A + B$, then $kC = kA + kB$. Of course, to form the difference $C = A - B$ of two matrices according to these rules, we simply multiply B by the scalar -1 and add the result to A; in this case $c_{ij} = a_{ij} - b_{ij}$, as expected.

An $(m \times n)$ matrix A and an $(n \times p)$ matrix B may be multiplied to form an $(m \times p)$ matrix $C = AB$ by the rule

$$c_{ij} = \sum_{r=1}^{n} a_{ir} b_{rj} \tag{2.2}$$

for $i = 1, 2, \ldots, m$ and $j = 1, 2, \ldots, p$. For example, the product of a (2×3) matrix and a (3×4) matrix gives a (2×4) matrix:

$$\begin{bmatrix} 1 & 0 & -1 \\ 2 & -1 & 1 \end{bmatrix} \cdot \begin{bmatrix} -1 & 1 & 0 & 1 \\ 2 & 1 & 3 & -1 \\ -1 & 2 & 1 & 1 \end{bmatrix} = \begin{bmatrix} 0 & -1 & -1 & 0 \\ -5 & 3 & -2 & 4 \end{bmatrix}.$$

Here we obtained c_{12}, for example, by taking the sum of the products of the elements in the first row of A (the first matrix) and the elements in the second column of B (the second matrix):

$$c_{12} = (1)(1) + (0)(1) + (-1)(2) = -1.$$

(This is the familiar rule for taking the dot or scalar product of two vectors in three-dimensional space.) For multiplication to be defined, the first matrix in the product must have the same number of columns as the second matrix has rows. In the preceding example, BA would not even be defined, since B is a (3×4) matrix and A is a (2×3) matrix, and $4 \neq 2$. As this example shows, we need *not* have equality between AB and BA, so that, in general, *matrix multiplication need not be commutative*. In fact, even in A and B are square matrices of the same order, we frequently find that $AB \neq BA$. A single example suffices:

$$\begin{bmatrix} 1 & 2 \\ -1 & 1 \end{bmatrix} \cdot \begin{bmatrix} 3 & 1 \\ 0 & -1 \end{bmatrix} = \begin{bmatrix} 3 & -1 \\ -3 & -2 \end{bmatrix},$$

$$\begin{bmatrix} 3 & 1 \\ 0 & -1 \end{bmatrix} \cdot \begin{bmatrix} 1 & 2 \\ -1 & 1 \end{bmatrix} = \begin{bmatrix} 2 & 7 \\ 1 & -1 \end{bmatrix}.$$

However, multiplication is *associative*. Let A be $(m \times n)$, B be $(n \times p)$, and C be $(p \times r)$. Then it turns out not to matter whether we first form AB and then multiply AB by C to form $(AB)\,C$, or whether we first form BC and then $A(BC)$; that is,

$$(AB)\,C = A(BC)\ .$$

To prove this, observe that the (i, j)th element in the first product is given by

$$\sum_{s=1}^{p} \left(\sum_{k=1}^{n} a_{ik} b_{ks} \right) c_{sj}\ ,$$

and in the second product by

$$\sum_{k=1}^{n} a_{ik} \left(\sum_{s=1}^{p} b_{ks} c_{sj} \right).$$

These two double sums are the same, since each product term in the one also occurs in the other, and this completes the proof. It follows that we can write the product of three matrices as ABC, without specifying which pair is to be multiplied first (although, as a practical note, the number of computer operations to form ABC could depend markedly on this choice). Similarly, we would have $((AB)C)D = (AB)(CD) = A(B(CD))$, and so forth, so that, in general, the product of the n matrices $A_1, A_2, \ldots , A_n$ can be written unambiguously as $A_1 A_2 \cdots A_n$.

Note also that matrix multiplication is *distributive;* it follows directly from the definition that

$$B(A_1 + A_2 + \cdots + A_n) = BA_1 + BA_2 + \cdots + BA_n.$$

The *transpose* A^{T} of an $(m \times n)$ matrix A is defined to be the $(n \times m)$ matrix obtained by transposing, or interchanging, the rows and columns of A. Denoting the (i, j)th element of A^{T} by a_{ij}^{T}, we have $a_{ij}^{\mathrm{T}} = a_{ji}$. An example would be

$$A = \begin{bmatrix} 1 & 2 & 3 \\ 2 & 3 & 1 \end{bmatrix}, \qquad A^{\mathrm{T}} = \begin{bmatrix} 1 & 2 \\ 2 & 3 \\ 3 & 1 \end{bmatrix}.$$

An interesting result is that $(AB)^{\mathrm{T}} = B^{\mathrm{T}} A^{\mathrm{T}}$; that is, the transpose of a product is the product of the transposes in reverse order. The reader should prove this by using the definition of a product to write down the formula for the (i, j)th element on each side of the equation; as always, an example should be tried.

The above rules for matrix manipulation were originally motivated by the desire to write a linear equation set in compact form. Let (a_{ij}), for $i = 1, 2, \ldots , n$ and $j = 1, 2, \ldots , n$, be a given set of constants, and let $b_1, b_2, \ldots , b_n$ be another set of constants. One often encounters the problem of finding n quantities, $x_1, x_2, \ldots , x_n$, that satisfy the linear equation set

$$a_{11}x_1 + a_{12}x_2 + \cdots + a_{1n}x_n = b_1 ,$$

$$a_{21}x_1 + a_{22}x_2 + \cdots + a_{2n}x_n = b_2 ,$$

$$\vdots$$

$$a_{n1}x_1 + a_{n2}x_2 + \cdots + a_{nn}x_n = b_n . \tag{2.3}$$

If we denote the (a_{ij}) coefficient array by the $(n \times n)$ matrix A, and the (x_i) and (b_i) arrays by the $(n \times 1)$ matrices

$$x = \begin{bmatrix} x_1 \\ x_2 \\ \vdots \\ x_n \end{bmatrix} , \quad b = \begin{bmatrix} b_1 \\ b_2 \\ \vdots \\ b_n \end{bmatrix} , \tag{2.4}$$

then Equations (2.3) can be written very compactly as

$$Ax = b .$$

Column matrices, such as those occurring in Equations (2.4), are termed *column vectors*; the individual elements can be thought of as the components of a vector in *n*-dimensional space.

As in the initial examples, matrix elements may be complex numbers, which, of course, may be manipulated in accordance with the usual algebraic rules for real numbers. In most of this chapter, however, we will be primarily concerned with real elements.

2.2 SQUARE MATRIX; INVERSE MATRIX

In this section we consider square matrices only, and we begin with some special matrices of simple form. We say that D is a *diagonal matrix* if all elements d_{ij} for which $i \neq j$ are zero. Thus

$$\begin{bmatrix} 2 & 0 & 0 \\ 0 & 1 & 0 \\ 0 & 0 & -1 \end{bmatrix} \quad \text{and} \quad \begin{bmatrix} 0 & 0 & 0 \\ 0 & 1 & 0 \\ 0 & 0 & -1 \end{bmatrix}$$

are diagonal matrices. (Incidentally, in any square matrix A, the elements a_{ij} for which $i = j$ are said to constitute the *main diagonal* or *principal diagonal*.). A special case of a diagonal matrix is the unit matrix I, for which all main diagonal elements are unity. For example,

$$[1], \quad \begin{bmatrix} 1 & 0 \\ 0 & 1 \end{bmatrix}, \quad \begin{bmatrix} 1 & 0 & 0 \\ 0 & 1 & 0 \\ 0 & 0 & 1 \end{bmatrix}, \quad \begin{bmatrix} 1 & 0 & 0 & 0 \\ 0 & 1 & 0 & 0 \\ 0 & 0 & 1 & 0 \\ 0 & 0 & 0 & 1 \end{bmatrix}$$

are unit matrices. Note that, given any square matrix A, we have $AI = IA = A$ (of course, the order of I must be the same as the order of A). This last equation exemplifies the fact that for *some* matrices, multiplication *is* commutative.

Given a square matrix A of order n, let us try to find a matrix B (square, of order n) such that

$$AB = I . \tag{2.5}$$

Such a matrix would naturally be termed a *right inverse* of A. Now in Equation (2.5), each column of I is obtained by multiplying the corresponding column of B by the matrix A (this statement is equivalent to the rule of Equation (2.2) for matrix multiplication), so Equation (2.5) really says

$$A \cdot \begin{bmatrix} b_{11} \\ b_{21} \\ \vdots \\ b_{n1} \end{bmatrix} = \begin{bmatrix} 1 \\ 0 \\ 0 \\ \vdots \\ 0 \end{bmatrix}, \quad A \cdot \begin{bmatrix} b_{12} \\ b_{22} \\ \vdots \\ b_{n2} \end{bmatrix} = \begin{bmatrix} 0 \\ 1 \\ 0 \\ \vdots \\ 0 \end{bmatrix},$$

$$\ldots, \quad A \cdot \begin{bmatrix} b_{1n} \\ b_{2n} \\ \vdots \\ b_{nn} \end{bmatrix} = \begin{bmatrix} 0 \\ 0 \\ \vdots \\ 1 \end{bmatrix} .$$

Thus, each column in the B matrix satisfies a certain linear equation set. But from determinant theory (reviewed in Appendix B), each of the equation sets has a unique solution if A is *nonsingular*—that is, if the determinant det (A) is nonzero. Consequently, if A is nonsingular, a unique right inverse B exists such that $AB = I$. We can go further, for if det (A) is nonzero, so is det (A^T) (since the two determinants are equal); consequently, we can also find a matrix C such that $A^T C = I$. Taking transposes, we find $(A^T C)^T = I^T$, or $C^T A = I$ (using the Sec. 2.1 rule for transposes). Thus, if det $(A) \neq 0$, A also has a left inverse. But multiplying $AB = I$

by C^T, we get $C^T AB = C^T I = C^T$; alternatively, in the lefthand matrix product, we can use $C^T A = I$. Thus, $IB = C^T$, or $B = C^T$, and it follows that the two inverses are identical. We denote the inverse of A by A^{-1}; we now have the result that *if A is nonsingular, a unique matrix A^{-1} exists such that $A^{-1}A = AA^{-1} = I$.* If A and B are nonsingular ($n \times n$) matrices, it follows at once that $(AB)^{-1} = B^{-1}A^{-1}$, since

$$(B^{-1}A^{-1})(AB) = B^{-1}(A^{-1}A)\, B = B^{-1}B = I.$$

If we know the inverse matrix A^{-1} for a given square matrix A, then we can write down formally the solution of a linear equation set of the form $Ax = b$. If each side is multiplied by A^{-1}, we obtain

$$A^{-1}Ax = A^{-1}b ,$$

and since $A^{-1}A = I$ and $Ix = x$, this reduces to $x = A^{-1}b$. Since A^{-1} is unique, this is also the only possible solution. Actually, it is laborious to compute A^{-1}, and in practice one should almost never do it. We will discuss a more efficient method for solving $Ax = b$ in Section 2.3.

A square matrix A is *symmetric* if $a_{ij} = a_{ji}$ for all choices of i and j; equivalently, the condition is that $A = A^T$. The matrix A is *skew-symmetric* if $a_{ij} = -a_{ji}$ for all i, j; equivalently, $A = -A^T$. A skew-symmetric matrix must have only zeros on the main diagonal.

If all the elements of a square matrix A lying above the main diagonal vanish, then A is said to be a *lower triangular* matrix. Similarly, if $a_{ij} = 0$ for $i > j$, we say A is of *upper triangular* form. Observe that the product of two triangular matrices of the same kind is again a triangular matrix of the same kind. The determinant of a triangular matrix equals the product of the main diagonal elements, and so a triangular matrix is nonsingular if and only if each main diagonal element is nonzero. To find the inverse of a nonsingular triangular matrix is easy; for example, let

$$\begin{bmatrix} a_{11} & 0 & 0 \\ a_{21} & a_{22} & 0 \\ a_{31} & a_{32} & a_{33} \end{bmatrix} \cdot \begin{bmatrix} b_{11} & b_{12} & b_{13} \\ b_{21} & b_{22} & b_{23} \\ b_{31} & b_{32} & b_{33} \end{bmatrix} = \begin{bmatrix} 1 & 0 & 0 \\ 0 & 1 & 0 \\ 0 & 0 & 1 \end{bmatrix}.$$

Then considering the products in which only the elements in the first row of A participate, we find at once that $b_{11} = 1/a_{11}$, $b_{12} = 0$, $b_{13} = 0$. Using these results and moving to the second row of A, we get $a_{21}b_{11} + a_{22}b_{21} = 0$, which now determines b_{21}; the other two equations are $a_{22}b_{22} = 1$ and $a_{22}b_{23} = 0$, so $b_{22} = 1/a_{22}$ and $b_{23} = 0$. The products involving the third row of A similarly determine b_{31}, b_{32}, and b_{33}. Observe that A^{-1} is again lower triangular. In the same way the reader should verify that the inverse of a nonsingular upper triangular matrix is also of upper triangular form.

2.3 GAUSSIAN ELIMINATION

This is a very practical (and also very natural) method for solving a set of linear equations of the form $Ax = b$. The idea is to use the first equation of the set (2.3) to solve for x_1 in terms of $x_2, x_3, \cdots, x_n$ and to substitute the result in each remaining equation. Putting aside the first equation, there are then $n - 1$ equations in the remaining variables, and the process is now repeated by solving for x_2 in the first equation of the new set. Continuing, we eventually get an equation involving x_n alone, which determines x_n. The immediate preceding equation, which involved x_{n-1} and x_n only, now allows us to solve for x_{n-1}; this back substitution process can be continued until all of the x_i are obtained. As an example, consider

$$2x_1 - x_2 + x_3 = 1 \, ,$$

$$x_1 - 2x_2 - x_3 = 0 \, ,$$

$$4x_1 + 3x_2 + x_3 = -1 \, . \tag{2.6}$$

From the first equation, we find

$$x_1 = \tfrac{1}{2} x_2 - \tfrac{1}{2} x_3 + \tfrac{1}{2} \, ; \tag{2.7}$$

using this value of x_1 in the next two equations gives

$$-\tfrac{3}{2} x_2 - \tfrac{3}{2} x_3 = -\tfrac{1}{2} \, ,$$

$$5x_2 - x_3 = -3 \, . \tag{2.8}$$

The first equation of this new set gives

$$x_2 = -x_3 + \tfrac{1}{3} \, , \tag{2.9}$$

so that the second equation of (2.8) can be rewritten as

$$-6x_3 = -\tfrac{14}{3} \, . \tag{2.10}$$

Solving Equations (2.10), (2.9), and (2.7) in that order, we now find

$$x_3 = \tfrac{7}{9} \, , \qquad x_2 = -\tfrac{4}{9} \, , \qquad x_1 = -\tfrac{1}{9} \, .$$

As a check, these values can be substituted into Equations (2.6).

If Equations (2.6) are written in the form $Ax = b$, then the above manipulations involved only the elements of A and b. For the coefficient matrix A, we start with

$$A = \begin{bmatrix} 2 & -1 & 1 \\ 1 & -2 & -1 \\ 4 & 3 & 1 \end{bmatrix}$$

and then first add $-\frac{1}{2}$ times the first row to the second row, and next add -2 times the first row to the third row, to obtain

$$A' = \begin{bmatrix} 2 & -1 & 1 \\ 0 & -\frac{3}{2} & -\frac{3}{2} \\ 0 & 5 & -1 \end{bmatrix}.$$

These operations (together with similar ones on b, to be described shortly) are entirely equivalent to the elimination of x_1 from the second pair of equations. The a_{11} element, 2, in this process is termed the *pivot*. We next use the pivot $a'_{22} = -\frac{3}{2}$, and add $-5/(-\frac{3}{2})$, or $\frac{10}{3}$, times the second row to the third row to obtain finally

$$A'' = \begin{bmatrix} 2 & -1 & 1 \\ 0 & -\frac{3}{2} & -\frac{3}{2} \\ 0 & 0 & -6 \end{bmatrix}.$$

Exactly the same operations are, of course, applied to the column vector b by using the same multipliers; we obtain

$$b = \begin{bmatrix} 1 \\ 0 \\ -1 \end{bmatrix}, \qquad b' = \begin{bmatrix} 1 \\ -\frac{1}{2} \\ -3 \end{bmatrix}, \qquad b'' = \begin{bmatrix} 1 \\ -\frac{1}{2} \\ -\frac{14}{3} \end{bmatrix}.$$

After carrying out these matrix operations, we end up with the modified equation set

$$\begin{bmatrix} 2 & -1 & 1 \\ 0 & -\frac{3}{2} & -\frac{3}{2} \\ 0 & 0 & -6 \end{bmatrix} \cdot \begin{bmatrix} x_1 \\ x_2 \\ x_3 \end{bmatrix} = \begin{bmatrix} 1 \\ -\frac{1}{2} \\ -\frac{14}{3} \end{bmatrix},$$

from which we obtain first x_3, then x_2, and finally x_1 just as before. We have, in effect, reduced A to upper triangular form.

This process fails if one of the pivots is zero or very close to zero. Suppose, for example, that the matrix A' of the last paragraph happened to be

$$A' = \begin{bmatrix} 2 & -1 & 1 \\ 0 & 0 & -\frac{3}{2} \\ 0 & 5 & -1 \end{bmatrix}.$$

Then $a'_{22} = 0$ is certainly not a suitable pivot. However, we can write our equations in any order, and all we need do is interchange rows 2 and 3 in A' (and of course also in b') to obtain the nonzero pivot 5. In this example, we do not have to proceed any further, since the new A' would already be in upper triangular form; but in the general case of an $(n \times n)$ matrix, we would carry on with the new pivot as before.

If a multiple of one row is added to another, the determinant of the matrix is unchanged (Appendix B). Also, if two rows are interchanged, the effect on the determinant is to reverse its sign. Thus, if a matrix A is reduced to upper triangular form by these operations, the product of the final pivot elements (which is the determinant of the triangular matrix) must equal $\pm \det(A)$. If follows that if A is nonsingular, we must always be able to find nonzero pivot elements, using row interchanges as necessary.

Even if a particular pivot is nonzero, it may be so small in magnitude that we would end up adding very large multiples of its row to succeeding rows; the information originally contained in these succeeding rows would then affect only the last few significant figures of the new rows, and we can expect a loss in accuracy. To avoid this, it is common practice to interchange rows as necessary in order to obtain that pivot element, among all possible ones in the particular column being examined, whose absolute value is greatest. This process is termed *partial pivoting*. It is less common to use *full pivoting*, in which columns also may be interchanged (i.e., the x_i quantities are renumbered), to choose the largest possible new pivot, in absolute value, considering all remaining rows and columns.

It could also happen that a coefficient matrix is badly out of scale, in the sense that some or all of the elements in a particular row are much greater or much smaller in magnitude than the elements in the other rows; it is then intuitively reasonable to multiply each term of the corresponding row in the equation set by an appropriate constant. A similar remark holds for columns; in this case we would, in effect, be rescaling the corresponding x_i variable.

We have remarked that a small pivot—or, equivalently, a small value for det (A)—can lead to inaccuracy. This statement needs some elaboration, since the natural question, What is small? arises. After all, we could multiply each term of the equation set by some large constant to make any initially small (but nonzero) coefficient determinant as large as we please. A better measure of sensitivity is obtained by asking, If small relative changes are made in some of the elements of A or b (say $\frac{1}{10}$ of 1 %), what are the corresponding relative changes in the solution vector x? The ratio of the latter to the former, in a worst-case scenario, is called the *condition number* of a matrix. (See Prob. 2.14.) Some linear equation programs provide an estimate of the condition number; a rule of thumb is that if the condition number is 10^k, then the last k significant figures in the x-values are dubious (as a result of roundoff errors). Another method for investigating sensitivity is to repeat the whole calculation by using double precision; still another is to make random, slight alterations in A or b and observe the resulting change in x.

If a linear equation set $Ax = b$ is found to be very sensitive to small relative changes in A or b, in the sense that the corresponding relative change in x is large, we say that A is *ill-conditioned*.

2.4 LINEAR EQUATION SUBROUTINE

The following program uses Gaussian elimination to solve a set of linear algebraic equations. It is assumed that the coefficient matrix is not singular and that the equation set is reasonably scaled. It deliberately does not use partial pivoting (which the reader is asked, in Prob. 2.10, to insert; also, it has been written for clarity rather than optimal efficiency.

```
C   SUBROUTINE LINEQ(N,A,B,X) USES ELIMINATION TO SOLVE N LINEAR
C   EQUATIONS OF FORM AX = B.   INPUT PARAMETERS ARE ORDER N, MATRIX A,
C   AND VECTOR B.   OUTPUT IS SOLUTION VECTOR X.   NEITHER PIVOTING NOR
C   SCALING IS USED.   A AND B ARE ALTERED DURING ELIMINATION PROCESS;
C   PIVOTS APPEAR ON MAIN DIAGONAL OF MODIFIED MATRIX A.
C
C   INPUT:
C     N = NUMBER OF EQUATIONS
C     A = COEFFICIENT MATRIX   ( DIMENSION (N,N) )
C     B = RIGHT HAND SIDE VECTOR   ( DIMENSION N )
C
C   OUTPUT:
C     X = SOLUTION VECTOR   ( DIMENSION N )
C
        SUBROUTINE LINEQ(N,A,B,X)
        DIMENSION A(N,N),B(N),X(N)
        NM=N-1
        DO 6 I=1,NM
        IP=I+1
        R=A(I,I)
        DO 7 J=IP,N
        T=A(J,I)/R
C   THE NEW VARIABLE T IS USED TO AVOID REPEATED ARRAY ELEMENT CALLS
        B(J)=B(J)-T*B(I)
        DO 8 K=I,N
8       A(J,K)=A(J,K)-T*A(I,K)
7       CONTINUE
6       CONTINUE
C   NOW THAT A IS IN UPPER TRIANGULAR FORM, SOLVE FOR X
        DO 9 I=N,1,-1
        IP=I+1
        S=0.
        DO 10 J=IP,N
10      S=S+X(J)*A(I,J)
9       X(I)=(B(I)-S)/A(I,I)
        END
```

2.5 EXAMPLE OF A LINEAR EQUATION SET

A simple example is afforded by the Wheatstone bridge electrical circuit sketched in Figure 2.1. A battery of voltage E and internal resistance R_6 is connected across a circuit made up of the five resistors R_1, R_2, R_3, R_4, and R_5, as shown, and it is desired to determine the branch currents i_1, i_2, i_3, i_4, i_5, and i_6, for which positive senses have been chosen arbitrarily as shown. (A negative value for a particular branch current would indicate that it flows in the opposite direction.)

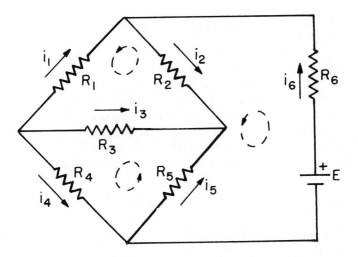

Figure 2.1. Wheatstone Bridge.

From physics, the conditions that these currents must satisfy (Kirchhoff's laws) are the net current flowing into any junction or node must be zero, and the net change in potential around any closed circuit must vanish.

The first of these conditions leads to

$$i_1 + i_6 - i_2 = 0 \, ,$$

$$-i_1 - i_3 - i_4 = 0 \, ,$$

$$i_2 + i_3 + i_5 = 0 \, ,$$

$$i_4 - i_5 - i_6 = 0 \, . \tag{2.11}$$

However, these equations are not all independent; if the first is solved for i_6, the second for i_4, the third for i_5, and the results substituted into the last equation, we will find that it reduces to $0 = 0$. So we discard this last equation, and use the first three only—thinking of them as determining, say, i_4, i_5, and i_6 in terms of i_1, i_2, i_3. This means that we have essentially three unknowns, i_1, i_2, and i_3, and we will need to apply to three independent circuits the condition of zero potential change around each. We choose the circuits indicated by dotted loops and obtain for the net potential increases:

$$i_1 R_1 - i_3 R_3 + i_2 R_2 = 0 \, ,$$

$$i_3 R_3 - i_4 R_4 - i_5 R_5 = 0 \, ,$$

$$E - i_2 R_2 + i_5 R_5 - i_6 R_6 = 0 \, . \tag{2.12}$$

These equations, together with the first three equations of (2.11), provide six equations in the six unknowns $i_1, \ldots, i_6$.

Of course, we could have made life a little easier by using the nodal current condition to reduce at once the number of unknowns, since, for example, each of i_4, i_5, i_6 can be written down at once in terms of i_1, i_2, i_3 (see Fig. 2.2) by simply looking at the nodes. The general rule is that any time we can't determine a particular current in terms of previously defined currents, introduce a new unknown. Notice that different choices were made for the positive senses for some of the branch currents in Figure 2.2. This is immaterial and simply a matter of convenience. The condition of zero potential change for the same three closed loops as before now gives

$$i_1 R_1 - i_3 R_3 + i_2 R_2 = 0 ,$$

$$i_3 R_3 + (i_1 + i_3) R_4 + (i_2 + i_3) R_5 = 0 ,$$

$$E - i_2 R_2 - (i_2 + i_3) R_5 + (i_1 - i_2) R_6 = 0 . \qquad \textbf{(2.13)}$$

This is a set of three, rather than six, equations. In a large problem it is usually worthwhile to reduce the number of variables—to the extent it is easy to do so—prior to formal use of a linear equation solver, because the number of computer operations required by a Gaussian elimination routine is proportional to the *cube* of the number of unknowns (cf. Prob. 2.5). Here the reduction process was suggested by the form of the physical problem itself.

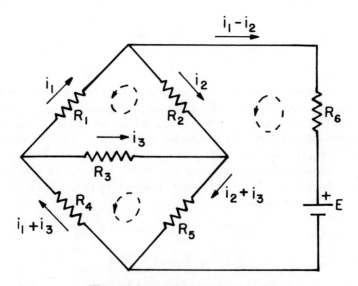

Figure 2.2. Modified branch currents.

Problems 2.6 and 2.8 ask the reader to apply the program of Section 2.4 to the equation set (2.13), for given parameter values, and also to a second electrical circuit problem.

It is frequently the case, in electrical circuit problems and elsewhere, that the coefficient matrix A in the linear equation set $Ax = b$ is *sparse*, in the sense that most of its elements are zero. Problem 2.15 describes one method for compact storage of a sparse matrix as well as an efficient technique for matrix multiplication. However, if $Ax = b$ is solved by a conventional elimination process, the modifications made to A will destroy its sparseness. Special techniques for dealing with sparse matrices are described in works by Dongarra and co-workers (1979) and Jennings (1977). See also the very important special case of a tridiagonal matrix considered in Problem 2.13.

2.6 ITERATIVE METHODS FOR $Ax = b$

The idea here is to make an initial guess for the solution vector x and then to correct it iteratively. Consider first an example for the set of three equations given by

$$2x_1 - x_2 + x_3 = 1 ,$$

$$-x_1 + 2x_2 - x_3 = 0 ,$$

$$x_1 + 2x_2 + 5x_3 = -1 . \qquad (2.14)$$

This set can be written as

$$2x_1 = 1 + x_2 - x_3 ,$$

$$2x_2 = 0 + x_1 + x_3 ,$$

$$5x_3 = -1 - x_1 - 2x_2 . \qquad (2.15)$$

The iteration procedure suggested by this form is to obtain a sequence of approximation vectors $x^{(p)}$, for $p = 1, 2, \cdots$, satisfying

$$2x_1^{(p+1)} = 1 + x_2^{(p)} - x_3^{(p)} ,$$
$$2x_2^{(p+1)} = 0 + x_1^{(p)} + x_3^{(p)} ,$$
$$5x_3^{(p+1)} = -1 - x_1^{(p)} - 2x_2^{(p)} . \qquad (2.16)$$

Suppose we try this out. As an initial guess, set $x_1^{(1)} = 1, x_2^{(1)} = 1, x_3^{(1)} = 1$. Then Equations (2.16) give the second approximation $x_1^{(2)} = \frac{1}{2}, x_2^{(2)} = 1, x_3^{(2)} = -.8$, and continuing, we find, for example, $x_1^{(6)} = .8363, x_2^{(6)} = .195, x_3^{(6)} = -.448$; $x_1^{(11)} = .8132, x_2^{(11)} = .1877, x_3^{(11)} = -.4380$; $x_1^{(16)} = .8125, x_2^{(16)} = .1875$,

$x_3^{(16)} = -.4375$. Thus after 15 iterations we have agreement to four significant figures with the exact solution $x_1 = .8125$, $x_2 = .1875$, $x_3 = -.4375$.

For this example, iteration is not competitive with successive elimination. However, as the number n of equations increases, the computational labor for the elimination process is proportional to the cube of n, and it does in fact turn out that for some large systems the most efficient method is a variant of this kind of iteration process.

The iteration method can be improved. First of all, the second equation of (2.16) does not make use of the (presumably) better value of x_1 obtained from the first equation, and it would certainly seem reasonable to do this. Following this idea, we are led to the revised process:

$$2x_1^{(p+1)} = 1 + x_2^{(p)} - x_3^{(p)} ,$$
$$2x_2^{(p+1)} = 0 + x_1^{(p+1)} + x_3^{(p)} ,$$
$$5x_3^{(p+1)} = -1 - x_1^{(p+1)} - 2x_2^{(p+1)} . \tag{2.17}$$

Starting again with $x^{(1)} = (1, 1, 1)$, the reader should check that we now obtain 4-figure solution accuracy after eight iterations—that is, the revised method is about twice as fast.

The method of Equation (2.16) is termed the *Jacobi* method, and that of Equation (2.17) is termed the *Gauss–Seidel* method (although it has been remarked that Gauss did not use it and Seidel disapproved of it!). Returning to the case of n linear equations $Ax = b$, write $A = L + D + U$, where the triangular matrix L contains all the terms of A lying below the main diagonal, D contains all of the diagonal terms, and U contains all of the terms above the main diagonal. Thus, in the example of Equations (2.14), we have

$$\begin{bmatrix} 2 & -1 & 1 \\ -1 & 2 & -1 \\ 1 & 2 & 5 \end{bmatrix} = \begin{bmatrix} 0 & 0 & 0 \\ -1 & 0 & 0 \\ 1 & 2 & 0 \end{bmatrix} + \begin{bmatrix} 2 & 0 & 0 \\ 0 & 2 & 0 \\ 0 & 0 & 5 \end{bmatrix} + \begin{bmatrix} 0 & -1 & 1 \\ 0 & 0 & -1 \\ 0 & 0 & 0 \end{bmatrix} .$$

Then the Jacobi method can be described by

$$Dx^{(p+1)} = b - (L + U)x^{(p)} , \tag{2.18}$$

and the Gauss–Seidel method by

$$(L + D)x^{(p+1)} = b - Ux^{(p)} . \tag{2.19}$$

Other splittings of A are also possible. We can write $A = S + (A - S)$ and use the iteration process

$$Sx^{(p+1)} = b - (A - S)x^{(p)} . \tag{2.20}$$

We would, of course, want to choose S so as to be easily invertible, so that the determination of $x^{(p+1)}$ from $x^{(p)}$ is immediate—as it is in Equations (2.18) and (2.19).

We have not yet discussed convergence, and in fact if this iteration method is applied to the problem of Equations (2.6), it will not converge. For the general case of Equation (2.20), suppose that at any stage the current iteration value $x^{(p)}$ differs from the true solution x by the error vector $\epsilon^{(p)}$, so that

$$x^{(p)} = x + \epsilon^{(p)} . \tag{2.21}$$

Subtracting $Sx = b - (A - S)x$ from Equation (2.20), we obtain

$$S\epsilon^{(p+1)} = (S - A)\epsilon^{(p)}$$

or

$$\epsilon^{(p+1)} = M\epsilon^{(p)} , \tag{2.22}$$

where $M = S^{-1}(S - A)$. Thus, starting with $p = 1$ and applying Equation (2.22) recursively, we have $\epsilon^{(p)} = M^{p-1}\epsilon^{(1)}$. Now $\epsilon^{(1)}$ depends on the initial guess and so can be almost any vector. Thus, for the process to converge, we need $M^{p-1}\epsilon^{(1)} \to 0$ as $p \to \infty$, for any arbitrary vector $\epsilon^{(1)}$. This can only happen if all elements of M^{p-1} approach zero as $p \to \infty$, and this is the condition for convergence. This condition has been used to verify the convergence of appropriate iterative methods for many of the linear equation sets that occur in practice; each case usually requires separate study.

For the present we note only that weighting can be used to improve convergence, just as in the use of an iterative method for a nonlinear equation. Instead of using $x^{(p+1)}$ as obtained from Equation (2.20),

$$x^{(p+1)} = S^{-1}b - S^{-1}(A - S)x^{(p)} , \tag{2.23}$$

we can form a weighted sum of this quantity with the previous iteration vector $x^{(p)}$, so that, as a revision of Equation (2.23), we use

$$x^{(p+1)} = (1 - r)x^{(p)} + r[S^{-1}b - S^{-1}(A - S)x^{(p)}] , \tag{2.24}$$

where the constant r represents the weighting factor. Equation (2.24) may also be written

$$x^{(p+1)} = x^{(p)} + r[\{S^{-1}b - S^{-1}(A - S)x^{(p)}\} - x^{(p)}] ,$$

where the quantity in brackets is the amount by which $x^{(p)}$ would change if the *relaxation parameter* r were set equal to unity. In solving linear equation sets that arise from the replacement of certain partial differential equations by finite difference equations, we will later encounter situations in which an optimal convergence rate is achieved for choices of $r > 1$; this *overrelaxation* advantage was noticed experimentally before the days of electronic computers.

2.7. MATRIX EIGENVALUES

Let a vector x in two-dimensional space have the two components x_1 and x_2, and consider the result obtained by multiplying it by a (2×2) matrix A:

$$y = Ax$$

or

$$\begin{bmatrix} y_1 \\ y_2 \end{bmatrix} = \begin{bmatrix} a_{11} & a_{12} \\ a_{21} & a_{22} \end{bmatrix} \begin{bmatrix} x_1 \\ x_2 \end{bmatrix}. \tag{2.25}$$

Corresponding to any chosen vector x, this process generates a new vector y in two-dimensional space and can be thought of as a transformation of vectors or as a mapping of vectors. With every original vector x is associated a new vector y, which usually will have a new magnitude and a new direction. A rather natural question now arises: Can we find a particular vector, z say, such that the vector Az resulting from the transformation has exactly the same direction as z? Geometrically, z may well get altered in length by the mapping, but it must not get turned. If we denote the ratio of the new length to the old length by λ, the requirement is that

$$Az = \lambda z , \tag{2.26}$$

where λ is some constant. If such a vector z exists, it is termed an *eigenvector* (from the German word *eigen*, meaning *proper* or *characteristic*), and the corresponding value of λ is termed an *eigenvalue*.

In terms of the components z_1 and z_2 of z, Equation (2.26) may be written

$$(a_{11} - \lambda)z_1 + a_{12}z_2 = 0 ,$$

$$a_{21}z_1 + (a_{22} - \lambda)z_2 = 0 . \tag{2.27}$$

But this is a pair of linear "homogeneous" equations in z_1 and z_2, in the sense that the righthand sides vanish. By Cramer's rule the only possible solution is given by $z_1 = 0$, $z_2 = 0$, unless the determinant of the coefficients vanishes. Thus, the only hope for finding a nontrivial solution to Equations (2.27) is to require λ to be one of the two roots of the quadratic equation obtained from $\det (A - \lambda I) = 0$:

$$\begin{vmatrix} a_{11} - \lambda & a_{12} \\ a_{21} & a_{22} - \lambda \end{vmatrix} = 0 \tag{2.28}$$

or

$$(a_{11} - \lambda) (a_{22} - \lambda) - a_{12}a_{21} = 0 .$$

Let λ_1 and λ_2 denote the two roots of this equation. In theoretical work we can usually require λ_1 and λ_2 to be different, since the case in which they are equal can be considered to be a limiting case of different λ's obtained by slightly altering one of the a_{ij} values.

If Equation (2.28) is satisfied, then this corresponds to one equation of (2.27) being a multiple of the other (remember that a determinant is unaltered if a multiple of one row is added to another; this fact can be used to prove easily the statement just made), so that we have in effect only one independent equation linking z_1 and z_2. One of z_1, z_2 may be given an arbitrary value, and the other is then determined. (There must be some arbitrariness, for if z is an eigenvector of Eq. (2.26), so is any constant times z.)

As an example, the eigenvalues and eigenvectors of the matrix

$$A = \begin{bmatrix} 2 & 1 \\ 1 & 2 \end{bmatrix}$$

are easily found to be

$$\lambda_1 = 3 , \quad z^{(1)} = \begin{bmatrix} 1 \\ 1 \end{bmatrix} ,$$

$$\lambda_2 = 1 , \quad z^{(2)} = \begin{bmatrix} 1 \\ -1 \end{bmatrix} .$$

Again, each of these two eigenvectors is only determined within an arbitrary multiplicative constant.

Next define a matrix E whose columns are made up of the two eigenvectors

$$E = \begin{bmatrix} 1 & 1 \\ 1 & -1 \end{bmatrix} .$$

Consider now the result of forming the product AE. This matrix multiplication can be thought of as being carried out on E column-by-column, and since the columns are just the eigenvectors, the first column $z^{(1)}$ gets replaced by $\lambda_1 z^{(1)}$, and the second by $\lambda_2 z^{(2)}$, so that

$$AE = \begin{bmatrix} \lambda_1 & \lambda_2 \\ \lambda_1 & -\lambda_2 \end{bmatrix} = \begin{bmatrix} 1 & 1 \\ 1 & -1 \end{bmatrix} \begin{bmatrix} \lambda_1 & 0 \\ 0 & \lambda_2 \end{bmatrix} .$$

In other words,

$$AE = ED , \tag{2.29}$$

where D is a diagonal matrix whose elements are the eigenvalues. Multiplying both sides of Equation (2.29) by E^{-1}, we obtain

$$E^{-1}AE = D . \tag{2.30}$$

Thus, the application of E^{-1} and E to A, in the manner of Equation (2.30), has diagonalized A; Equation (2.30) is said to constitute a *similarity transformation* of A. We will use this device in Chapter 6 in connection with the stability of differential equation algorithms.

More generally, let A be an $(n \times n)$ square matrix. If a nontrivial $(n \times 1)$ column vector z and a constant λ exist such that $Az = \lambda z$, then z is termed an eigenvector of A, and λ is termed an eigenvalue. Writing out the individual equation set, the possible values of λ must satisfy det $(A - \lambda I) = 0$, and there are n such λ_i values, obtainable as the roots of an nth-order polynomial equation* in λ. Again, a matrix E whose columns are the eigenvectors may be constructed, such that $E^{-1}AE = D$, where D is diagonal. (In the (2×2) case, E was clearly nonsingular; if the λ_i are all different, this also holds in the general case, although we omit the general proof, since the result can be verified directly in the particular applications made in this book.) It may be noted that some of the λ_i may well be complex numbers, but this does not affect the validity of the algebra.

ANNOTATED BIBLIOGRAPHY

J. J. Dongarra, C. B. Moler, J. R. Bunch, and G. W. Stewart, 1979, *LINPACK User's Guide*, Society for Industrial and Applied Mathematics, Philadelphia, 363p.

LINPACK is a carefully constructed package of computer subroutines that solve sets of simultaneous linear algebraic equations. The User's Guide describes the rationale behind each routine, the input and output conventions, and precautions; it also gives program listings. This Guide is a textbook in itself and is well worth its modest price. An interesting remark on page 1.34, in connection with Gaussian elimination, is that an example occurs in a 2000-year-old Chinese manuscript.

A. Jennings, 1977, *Matrix Computation for Engineers and Scientists*, John Wiley, New York, 330p.

This compact but instructive book is not as well known as it should be. The emphasis is on practical application and computer technique.

B. Noble and J. W. Daniel, 1977, *Applied Linear Algebra*, 2nd ed., Prentice-Hall, Englewood Cliffs, New Jersey, 477p.

The treatment here is also very clear. Many examples (transportation, frameworks, electric circuits, and even beetle cycles) are given of practical problems that lead to linear equation sets.

B. T. Smith, J. M. Boyle, J. J. Dongarra, B. S. Garbow, Y. Ikebe, V. C. Klema, and C. B. Moler, 1976, *Matrix Eigensystem Routines—EISPACK Guide*, 2nd ed., Springer-Verlag, New York, 551p.

This does for eigenvalues what LINPACK does for linear equations. It is, however, less tutorial and concentrates on program listings and user guidance.

G. Strang, 1976, *Linear Algebra and its Applications*, Academic Press, New York, 274p.

This text gives a very readable introduction to matrix operations and to linear equations. Chapter 1 covers the basics.

---

*For more sophisticated methods of finding the λ_i, see works by Strang (1976), Noble and Daniel (1977), and Jennings (1977); see also Problem 2.16.

PROBLEMS

2.1 Let A and B be defined by

$$A = \begin{bmatrix} 0 & 1 & -1 \\ 2 & 1 & 1 \\ 0 & 1 & 3 \end{bmatrix}, \quad B = \begin{bmatrix} -1 & 0 & 2 \\ 1 & -1 & 1 \\ 2 & -1 & 2 \end{bmatrix}.$$

Verify that each is nonsingular, and find A^{-1}, B^{-1}. Check that $A^{-1}A = AA^{-1} = I$ and that $(AB)^{-1} = B^{-1}A^{-1}$.

2.2 Use determinant theory (Appendix B) to show that a singular square matrix cannot have an inverse. [*Hint:* Consider the determinant of the product of A and A^{-1}.]

2.3 Find x in $Ax = b$ if

$$A = \begin{bmatrix} 1 & 1 + \epsilon \\ 1 & 1 \end{bmatrix}, \quad b = \begin{bmatrix} 1 + \alpha \\ 1 \end{bmatrix},$$

where ϵ and α are very small compared to unity. Consider the cases $\alpha = \epsilon$, $\alpha = 0$, and observe the large relative change in x resulting from a small relative change in b.

2.4 Use the subroutine of Section 2.4 (and an appropriate main program) to solve Equations (2.6). Next, solve $Ax = b$ for an nth order system, in which $a_{ij} = \exp(-|i - j|)$ and $b_j = 1/j$, for all i, j values in the range $(1, n)$. Use $n = 10, 20, \cdots$, and plot computer time versus n^3.

2.5 (a) Assuming that addition or subtraction takes little time compared to multiplication or division, show that the number of computer arithmetical operations required to solve a set of n equations by elimination is about $\frac{1}{3} n^3$ if n is large. A very fast computer might do about 100 operations in a microsecond; estimate the time required to solve 10^2, 10^3, 10^4, and 10^5 equations. (b) Observe that the inversion of A is equivalent to the solution of the problems

$$Ax = \begin{bmatrix} 1 \\ 0 \\ 0 \\ \vdots \end{bmatrix}, \quad Ax = \begin{bmatrix} 0 \\ 1 \\ 0 \\ 0 \\ \vdots \end{bmatrix}, \quad \text{and so forth,}$$

and (noting also that we need manipulate A only once) estimate the number of operations required for matrix inversion. Compare with the number of operations required to multiply together two matrices of order n.

2.6 In Figure (2.2) let all resistors have a value of unity, with $E = 1$; write a main program to solve Equations (2.11) and (2.12), and, alternatively, Equations

(2.13), by use of the subroutine of Section (2.4). What features of the result provide a check? Repeat, using different R_i values of your choice.

2.7 Write a subroutine that forms the product Ax for given matrices A (order $n \times n$) and x (order $n \times 1$). Use this subroutine to check the result obtained in Problem 2.6. (The original A should be used, not the modified A produced by LINEQ.)

2.8 Solve for the currents in the ladder network of Figure 2.3 for $n = 10$, 20, 40, 80. [*Hint:* This problem is easier than it looks.]

2.9 When the author first wrote the LINEQ subroutine, he tried to make it as transparent as possible (even at the cost of repeatedly accessing the same matrix storage cell). Instead of introducing the variables R and T, the statement numbered 8 read

$$8 \ A(J, K) = A(J, K) - A(J, I)*A(I, K)/A(I, I) .$$

To his chagrin, he discovered that this subroutine gave entirely wrong results. Why?

2.10 Insert partial pivoting into LINEQ (see Sec. 2.4) and try it out.

2.11 In Section 2.3 the matrix A' could be obtained from A by the operation $A' = E_2E_1A$, and A'' from A' via $A'' = E_3A'$, where E_1, E_2, and E_3 are the lower triangular matrices.

$$E_1 = \begin{bmatrix} 1 & 0 & 0 \\ -\frac{1}{2} & 1 & 0 \\ 0 & 0 & 1 \end{bmatrix}, \quad E_2 = \begin{bmatrix} 1 & 0 & 0 \\ 0 & 1 & 0 \\ -2 & 0 & 1 \end{bmatrix}, \quad E_3 = \begin{bmatrix} 1 & 0 & 0 \\ 0 & 1 & 0 \\ 0 & \frac{10}{3} & 1 \end{bmatrix}.$$

Generalize this observation to deduce that in the Gaussian elimination process a matrix A is repeatedly multiplied by lower triangular matrices $\cdots E_3E_2E_1 = L_1$, say (where L_1 is also lower triangular) to produce the result $L_1A = U$ (provided that no pivot vanishes). Thus, with $L = L_1^{-1}$, we find $A = LU$, which is a decomposition into the product of a lower triangular matrix L, with unit elements on its main diagonal, and an upper triangular matrix U. (Note also that as long as A is nonsingular, rows may be appropriately interchanged to yield nonzero pivots; the permuted matrix could then be decomposed in this manner.) Finally, if the diagonal elements of U (the pivots) are made the elements of a diagonal matrix D, we have the decomposition $A = LDU_1$, where both L and U_1 now have unit

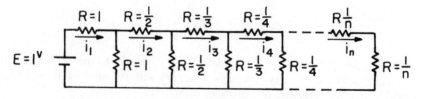

Figure 2.3. Circuit for Problem 2.8.

elements on their main diagonals. *Question:* How can L be economically constructed during Gaussian elimination? [*Hint:* Compute E_1^{-1} and note its relation to E_1, also $(E_2 E_1)^{-1}$, etc.]

2.12 Show that if a square matrix A has an *LDU* decomposition of the kind described in Problem 2.11, then that decomposition is unique. Show that if A is symmetric, then $L = U^T$, and if the pivots of A are also all positive, then the square roots of the these pivots can be put on the main diagonals of the L and U matrices to yield the so-called *Cholesky decomposition* $A = (L\sqrt{D})\,(L\sqrt{D})^T$.

2.13 A square matrix A is *tridiagonal* if all of its elements vanish, except perhaps those on the main diagonal and the two adjoining diagonals. That is, the nonvanishing elements a_{ij} must satisfy the condition $i - j = 0, +1$, or -1. (This is a case which often occurs in the numerical solution of differential equations.) It is clear that the elimination process becomes particularly simple for a tridiagonal matrix. In fact, the equations in the set $Ax = b$ may be written

$$a_{11}x_1 + a_{12}x_2 = b_1 \, ,$$

$$a_{21}x_1 + a_{22}x_2 + a_{23}x_3 = b_2 \, ,$$

$$a_{32}x_2 + a_{33}x_3 + a_{34}x_4 = b_3 \, ,$$

$$\vdots$$

$$a_{n, n-1}x_{n-1} + a_{nn}x_n = b_n \, . \tag{2.31}$$

We now write

$$x_i = c_i x_{i+1} + d_i \tag{2.32}$$

and observe that the c_i and d_i can be obtained in sequence, starting with the first Equation of (2.31). Proceeding step by step to the last Equation of (2.31), we find that this equation now allows us to determine x_n, so that Equation (2.32) can now be used (working backwards) to obtain all of the x_i.

Write a subroutine which uses this idea to solve $Ax = b$, where A is tridiagonal. How many operations are required if there are n equations?

Use your subroutine to solve the problem $Ax = b$, where

$$A = \begin{bmatrix} 1 & 0 & 0 & \cdots & & & \\ -1 & 2 & -1 & 0 & \cdots & & \\ 0 & -1 & 2 & -1 & 0 & \cdots & \\ 0 & 0 & -1 & 2 & -1 & 0 & \cdots \\ \cdots\cdots\cdots\cdots\cdots\cdots\cdots\cdots\cdots\cdots\cdots\cdots\cdots\cdots & & \\ & & & & -1 & 2 & -1 \\ & & & & 0 & 0 & 1 \end{bmatrix}$$

and where all $b_i = 0$, except that $b_1 = 3$ and $b_n = 7$. (Note that $c_1 = 0$.) Take $n = 20$, $n = 100$, $n = 1000$. What features of the result provide a check?

2.14 It is sometimes useful to have a measure of how large a square matrix A is via its *norm* $\|A\|$. A conventional approach is to first define the norm of a vector x by $\|x\| = (x_1^2 + x_2^2 + \cdots x_n^2)^{1/2}$, which is, of course, a generalization of the idea of the length of a three-dimensional vector. If some vector x is chosen, then Ax is a new vector that can be thought of as a transformation of x (or a mapping) corresponding to the application of the "operator" A. This new vector Ax has norm $\|Ax\|$, and we can now obtain a measure of the "size" of A by asking how large the new vector $\|Ax\|$ is compared to the original vector x. Considering all possible nonzero vectors x, we define

$$\|A\| = \max_{x \neq 0} \frac{\|Ax\|}{\|x\|} .$$

The *condition number c* can then be defined by

$$c = \|A\| \cdot \|A^{-1}\| .$$

Would c be altered if each element of A were multiplied by some constant? For the case of a (2×2) matrix A, discuss the reasonableness of this definition in terms of how x is affected by small changes in A or b.

Next, explain why, for any given vector y, $\|Ay\| \leq \|A\| \cdot \|y\|$. Let x be a solution of $Ax = b$, and suppose that b is altered by a small amount δb, with $A\delta x = \delta b$. Show that

$$\|b\| \leq \|A\| \cdot \|x\|, \qquad \|\delta x\| \leq \|A^{-1}\| \cdot \|\delta b\| ,$$

and deduce that

$$\frac{\|\delta x\|}{\|x\|} \leq c \frac{\|\delta b\|}{\|b\|} ,$$

so that the condition number provides a bound for the relative change in the solution vector x resulting from a small perturbation in b. (A somewhat similar result can be obtained if a small change δA in A is considered.) This kind of result explains why c is a measure of the sensitivity of the problem $Ax = b$ to small changes in either A or b.

2.15 It often happens that most of the individual equations, in a given set of linear equations, involve only a few of the variables. This is commonly the case in structural modeling, in large electrical networks, and in the replacement of partial differential equations by finite difference or finite element approximations (see Chapter 8). In such cases the coefficient matrix A is termed *sparse*; most of its elements vanish.

If A is very large, it can be useful to store only the nonzero terms. One way of doing this is to store three one-dimensional arrays, say A_1, A_2, and A_3. The array A_1 lists, in any order, the nonzero elements a_{ij} of A, and corresponding locations in A_2 and A_3 give the values of i and j for each such a_{ij}. For example, the arrays for

$$A = \begin{bmatrix} 0 & 0 & .5 \\ -1 & 0 & .1 \\ 0 & 0 & 0 \end{bmatrix}$$

would be

$$A_1 = (.5, -1, .1) , \quad A_2 = (1, 2, 2) , \quad A_3 = (3, 1, 3) .$$

A fourth array, A_4, could be included that gives the address of the next nonzero element on that row; this minimizes searching. See the book by Jennings (1977) for other ideas.

Suppose now that the product of two ($n \times n$) matrices A and B is to be formed to yield $C = AB$, where A (but not necessarily B) is sparse. Explain why the following program fragment is appropriate, and devise a similar technique for the operation $D = BA$.

```
      DO 6 I=1,N
      DO 7 J=1,N
7     C(I,J)=0.
      DO 6 K=1,N
      XX=A(I,K)
      IF(XX.EQ.0.) GO TO 6
      DO 8 J=1,N
      C(I,J)=B(K,J)*XX+C(I,J)
8     CONTINUE
6     CONTINUE
```

2.16 A method that can often be used to find the largest eigenvalue (in absolute value) and corresponding eigenvector for a matrix A is to start with an arbitrary vector ξ and form $A\xi$, $A^2\xi$, $A^3\xi$, $\cdots$, by repeated multiplication by A. If the eigenvalues $z^{(i)}$ are distinct, and if $|\lambda_1|$ is greater than all other $|\lambda_i|$, then $A^n\xi$ will approach a multiple of $\lambda_1^n z^{(1)}$. For the example

$$A = \begin{bmatrix} 2 & 1 \\ 1 & 2 \end{bmatrix}$$

of Section (2.7), prove that this will indeed be the case. [*Hint:* Write $\xi = \alpha z^{(1)} + \beta z^{(2)}$ and explain how the sequence of iterates can be used to determine λ_1 and $z^{(1)}$.] Experiment with this method, starting, say, with

$$\xi = \begin{bmatrix} 1 \\ 2 \end{bmatrix}.$$

2.17 Suppose that all of A, x, and b in $Ax = b$ are complex quantities. Rephrase the equation set so as to involve only real numbers; how does the size of the new system compare to the original size?

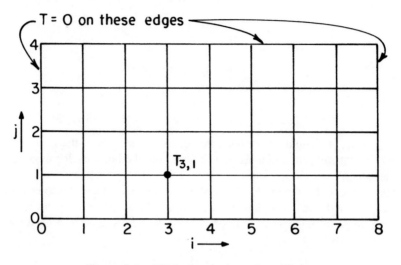

Figure 2.4. Thick plate in thermal equilibrium.

2.18 A thick rectangular plate, with insulated faces, is allowed to come to thermal equilibrium under the influence of a temperature distribution applied to its edges. Mesh points are chosen as shown in Figure 2.4, and the temperature at the location (i, j) is denoted by $T_{i,j}$. It is a property of the steady-state temperature field in a uniform medium that the temperature at any point is equal to the average of the temperatures over any circular area centered at that point; this suggests that, as an approximation, the mesh point temperature should satisfy the equation

$$T_{i,j} = \tfrac{1}{4}(T_{i,j+1} + T_{i,j-1} + T_{i+1,j} + T_{i-1,j}) .$$

Let the temperature along the vertical and top edges be zero (where zero is any chosen reference level), and let the temperature along the lower edge be such that an appropriate choice for mesh point values is $T_{i,0} = 4i(8 - i)$ for $i = 0, 1, \cdots , 8$. There are then 21 unknown temperatures and an equal number of equations. Solve this equation set for the unknown $T_{i,j}$, both by elimination and by iteration. Can you find a good relaxation factor?

3

SIMULTANEOUS NONLINEAR EQUATIONS

Numerical methods applicable to the solution of a single nonlinear equation were discussed in Chapter 1. Frequently, however, problems requiring the solution of two or more coupled nonlinear equations are encountered, and this topic provides the subject matter of the present chapter. To illustrate the way in which such a problem can arise, as well as some of the solution ideas, we begin with a pipe flow example of the kind met with in gas or water distribution networks.

3.1 PIPE FLOW EXAMPLE

Over a wide range of conditions, the pressure difference that must be applied across the two ends of a pipe in order to sustain a steady flow of fluid through that pipe is found by experiment to be proportional to the 7/4 power of the flow rate. Consider as an example the simple piping network of Figure 3.1 where three pipes, perhaps of different lengths, diameters, and degrees of internal roughness, meet at a common node. Pressures P_1 and P_2 are applied at inlet ports 1 and 2, and a pressure P_3 is maintained at the exit port 3. As a result, flow rates Q_1, Q_2, and $Q_1 + Q_2$ are obtained as shown in the figure. The pressure P_0 at the junction node is not specified; it adjusts itself as required. Denoting the constant of proportionality in the pressure drop law for the ith pipe section by C_i, we have

$$P_1 - P_0 = C_1 Q_1^{7/4} ,$$
$$P_2 - P_0 = C_2 Q_2^{7/4} ,$$
$$P_0 - P_3 = C_3 (Q_1 + Q_2)^{7/4} . \tag{3.1}$$

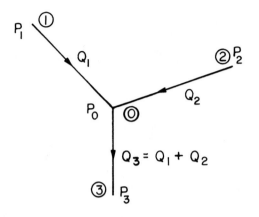

Figure 3.1 Piping network.

These equations are based on the assumption that the flow rates are in the directions shown. If the applied pressures are such that one or more of the flow directions alters, we would have to revise the equations accordingly. Alternatively, an equation like the first one of (3.1) could be written $P_1 - P_0 = C_1 Q_1 |Q_1|^{3/4}$, and this is valid for flow in either direction.

It is usually convenient to deal with equations in dimensionless form. Let P^*, Q^* be chosen reference values for pressure and flow rate, and define $p_i = P_i/P^*$, $q_i = Q_i/Q^*$, $c_i = C_i(Q^*)^{7/4}/P^*$. Equations (3.1) become

$$p_1 - p_0 = c_1 q_1^{7/4} ,$$

$$p_2 - p_0 = c_2 q_2^{7/4} ,$$

$$p_0 - p_3 = c_3(q_1 + q_2)^{7/4} . \tag{3.2}$$

This is a set of three nonlinear equations in the three unknowns p_0, q_1, q_2. One immediate possibility is to solve the first equation for p_0 in terms of q_1 and to use the result in the second equation so as to express q_2 in terms of q_1. Using now both results in the third equation, we obtain a single equation for q_1:

$$p_1 - p_3 - c_1 q_1^{7/4} = c_3[q_1 + c_2^{-4/7}(p_2 - p_1 + c_1 q_1^{7/4})^{4/7}]^{7/4} . \tag{3.3}$$

Equation (3.3) could now be solved by the methods of Chapter 1. This kind of elimination process is feasible here, where there are only three pipe sections, but in a practical problem there may be dozens or hundreds of sections, and if we start eliminating variables the equations soon become unmanageable (even if we arrange to have a computer keep track of the algebra).

Consequently, it is worthwhile to develop methods for the simultaneous solution of nonlinear equations, and in this chapter we consider a number of such methods. We will use Equations (3.2) as a prototype example, but before proceeding we might as well make one obvious simplification based on the fact that p_0 occurs linearly in these equations. Subtracting the second from the first, and then adding the third to the first, Equations (3.2) become

$$p_1 - p_2 = c_1 q_1^{7/4} - c_2 q_2^{7/4} \,,$$
$$p_1 - p_3 = c_1 q_1^{7/4} + c_3 (q_1 + q_2)^{7/4} \,. \qquad \textbf{(3.4)}$$

With p_1, p_2, p_3, c_1, c_2, and c_3 specified, this is now a set of two nonlinear equations in the two unknowns q_1 and q_2. Although these two equations will provide an example case for the methods of Sections (3.2) and (3.3), we should alert the reader that for more complicated networks than that of Figure 3.1, alternative and more efficient methods exist, which will be considered in Section 3.5.

3.2 SECANT PLANE METHOD

In general, the methods of Chapter 1 can be extended to sets of equations. Consider, for example, a coupled pair of equations of the form

$$f(x, y) = 0 \,, \qquad g(x, y) = 0 \,, \qquad \textbf{(3.5)}$$

where we want to find values of x and y for which both equations are simultaneously satisfied.

In applying the secant method of Section 1.3 to the problem $f(x) = 0$, the curve $y = f(x)$ was approximated locally by a straight line drawn through the two points (x_1, y_1) and (x_2, y_2), where $y_j = f(x_j)$ for $j = 1, 2$; the intersection of this line with the x-axis was then determined. The analogous procedure here is to choose three points in three-dimensional space and to pass a plane through these points. Define $z = f(x, y)$ and choose the points (x_1, y_1, z_1), (x_2, y_2, z_2), and (x_3, y_3, z_3), where $z_j = f(x_j, y_j)$ for $j = 1, 2, 3$. Pass a plane through these three points (cf. Fig. 3.2), and let the plane cut the (x, y) plane along a straight line A as shown. The line A is then an approximation to the true curve along which $f(x, y) = 0$. Repeat the process for the surface $z = g(x, y)$, obtaining now a plane that cuts the (x, y) plane along a straight line B (not shown). The intersection of A and B provides an approximation to the desired solution point (x, y). If we denote the coordinates of this intersection point by (x_4, y_4), then the process can be iterated, starting now, for example, with the three points (x_2, y_2, z_2), (x_3, y_3, z_3), and (x_4, y_4, z_4), where $z_i = f(x_i, y_i)$ for the first plane and $z_i = g(x_i, y_i)$ for the second.

For a numerical example, set $p_1 = 2$, $p_2 = 1.5$, $p_3 = 1$ in Equations (3.4), and take $c_1 = .75$, $c_2 = .0625$, $c_3 = .0275$. Equations (3.4) become

$$.5 = .75 q_1^{7/4} - .0625 q_2^{7/4} \,,$$
$$1 = .75 q_1^{7/4} + .0275 (q_1 + q_2)^{7/4} \,,$$

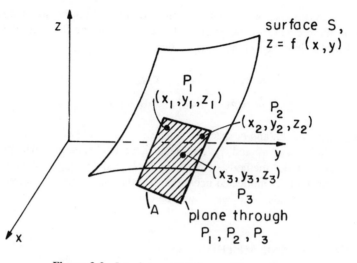

Figure 3.2 Local approximation to S by a plane.

or, with $x = q_1$ and $y = q_2$,

$$f(x, y) = 0 , \qquad g(x, y) = 0 ,$$

where

$$f(x, y) = .75x^{1.75} - .0625y^{1.75} - .5 ,$$

$$g(x, y) = .75x^{1.75} + .0275(x + y)^{1.75} - 1 . \tag{3.6}$$

A little experimentation suggests that $x \cong 1$ and $y \cong 2$, so we try $(x_1, y_1) =$ (1, 2), $(x_2, y_2) = (1.1, 2)$, $(x_3, y_3) = (1, 2.1)$. For $z = f(x, y)$, we find $z_1 =$ .03978, $z_2 = .17591$, $z_3 = .02104$. A plane through the three (x_i, y_i, z_i) points has the form $z = \alpha x + \beta y + \gamma$, where the constants α, β, γ are chosen so that the coordinates of these three points satisfy this equation. Setting $z = 0$, we then find that this plane cuts the (x, y) plane along the line $1.3612x - .1874y - .9466 =$ 0. Similarly, using the same three (x_i, y_i) points, but with $z = g(x, y)$, we obtain $z_1 = -.06194$, $z_2 = .08530$, $z_3 = -.05083$, and the plane defined by these three new points cuts the (x, y) plane along the line $1.4724x + .1111y - 1.7565 = 0$. These two lines in the (x, y) plane intersect at $x_4 = 1.0168$, $y_4 = 2.3346$. Using (x_2, y_2), (x_3, y_3), and (x_4, y_4) in the same way, we find $x_5 = 1.0174$, $y_5 = 2.3220$. Similarly, $x_6 = 1.0175$, $y_5 = 2.3225$, and the process seems to be converging.

However, if we continue a little further, we find that the arithmetic becomes indefinite. What happened is that we end up trying to pass a plane through three points that are very close together, and of course this is numerically precarious. In choosing the set of three points for each new step, it is better to choose the current best point and two new adjoining points at some distance from it. We need not pursue this example further (particularly since Problem 3.1 asks the reader to write an appropriate program), but it will indeed turn out that a suitably modified procedure converges to $x = 1.0175$, $y = 2.3225$ without encountering any indefinite arithmetic.

In choosing the three points (x_i, y_i) one should be careful that they are not close to being collinear, since the associated plane may then not be well defined. One should also guard against the possibility that the two lines in the (x, y) plane, whose intersection is to be found, may intersect at a very small angle; this could lead the iteration process astray.

The idea of this section can be generalized to apply to sets of equations involving more variables. The case of three variables will make the method clear. Let it be required to find x, y, and z such that

$$f(x, y, z) = 0 , \qquad g(x, y, z) = 0 , \qquad h(x, y, z) = 0 , \qquad (3.7)$$

where f, g, h are three given functions. Then we choose four points in four-dimensional space (x, y, z, w), denoted by (x_i, y_i, z_i, w_i), where $w_i = f(x_i, y_i, z_i)$ for $i = 1, 2, 3, 4$. We next pass a "hyperplane" of the form

$$\alpha x + \beta y + \gamma z + \delta = w$$

through these four points by the appropriate choice of constants α, β, γ, δ. This plane intersects three-dimensional (x, y, z) space in the plane $\alpha x + \beta y + \gamma z + \delta = 0$. A similar process is applied to the g and h functions, and we end up with three planes in (x, y, z) space whose intersection point provides an approximation to the desired solution.

A set of nonlinear equations, such as Equations (3.5) or (3.7), may have several solutions, so one should note the possibility that the method of this section, or any other method, may be converging to a solution other than the one desired. Also, the speed of any iterative procedure can be enhanced by an improvement in the initial guess, so some effort expended on finding a good starting point can be well worthwhile.

3.3 SUCCESSIVE SUBSTITUTION

Consider again Equations (3.5) and write them in the form

$$x = F(x, y) , \qquad y = G(x, y) . \qquad (3.8)$$

For example, for the f and g functions of Equations (3.6), we could write

$$x = \left[\frac{.5 + .0625y^{1.75}}{.75} \right]^{4/7} , \qquad y = \left[\frac{1 - .75x^{1.75}}{.0275} \right]^{4/7} - x . \qquad (3.9)$$

Starting now with initial guesses (x_1, y_1), a natural procedure is to repeatedly substitute in Equations (3.8) to obtain

$$x_2 = F(x_1, y_1) , \qquad y_2 = G(x_2, y_1) ,$$

and, in general,

$$x_{j+1} = F(x_j, y_j) , \qquad y_{j+1} = G(x_{j+1}, y_j) \qquad \textbf{(3.10)}$$

for $j = 1, 2, 3, \ldots$. Note that we have written $G(x_{j+1}, y_j)$ rather than $G(x_j, y_j)$ on the righthand side, so as to use the best current value for x; this choice may also enhance convergence. Because of the analogy to the single-variable case of Section 1.5, we can anticipate convergence of the iterative procedure if the derivatives of F and G with respect to each of x and y are sufficiently small.

For F and G as given by Equations (3.9), and with the same starting values as before (i.e., $x_1 = 1$ and $y_1 = 2$), the iterative process diverges. However, it can be rescued by the weighting device suggested in Section 1.5. Replacing Equations (3.10) by

$$x_{j+1} = Ax_j + (1 - A)F(x_j, y_j) ,$$

$$y_{j+1} = Ay_j + (1 - A)G(x_{j+1}, y_j) , \qquad \textbf{(3.11)}$$

and again using $x_1 = 1$, $y_1 = 2$, we see that the problem described by Equations (3.9) converges for A greater than about .2; the optimal choice is $A \cong .3$. In the general case the optimal choice for A need not lie between 0 and 1.

There is no difficulty in extending the iteration procedure to the case of n equations in n variables. The following subroutine provides one approach for the case $x_i = F_i(x_1, x_2, \ldots, x_n)$ for $i = 1, 2, \ldots, n$. The subroutine is followed by a main program suitable for the problem of Equations (3.9).

```
C   SUBROUTINE SUBST(N,NI,NIA,X,XX,FUN,TOL,STEP,A) SOLVES
C   SET OF N NONLINEAR EQUATIONS USING REPEATED SUBSTITUTION.
C   EQUATIONS MUST BE IN FORM
C           X(J) = FUN(J,X(1),X(2),...,X(N))
C   FOR J = 1,2,3,...,N.
C
C   INPUT:
C       N = NUMBER OF EQUATIONS
C      NI = MAXIMUM NUMBER OF ITERATIONS
C       X = ARRAY OF INITIAL VALUES, DIMENSION N
C      XX = WORKING ARRAY, DIMENSION N
C     TOL = TOLERANCE; SUBROUTINE EXITS IF STEP LENGTH
C               BECOMES LESS THAN TOL
C       A = WEIGHTING (PROPORTION OF PREVIOUS X VALUE
C               INCORPORATED IN NEW VALUE)
C
C   OUTPUT:
C       X = ARRAY CONTAINING CURRENT BEST VALUE
C     NIA = ACTUAL NUMBER OF ITERATIONS
C    STEP = LENGTH OF LAST ITERATION STEP
C
C   FUNCTION SUBPROGRAM CALLED:
C           FUN(J,X) GIVES VALUE OF NEW X(J).  MUST BE DECLARED
C               EXTERNAL BY CALLING PROGRAM
C
        SUBROUTINE SUBST(N,NI,NIA,X,XX,FUN,TOL,STEP,A)
        DIMENSION X(N),XX(N)
```

```
C    START ITERATING
         DO 5 II=1,NI
C    RECORD CURRENT X VALUES
         DO 6 I=1,N
6        XX(I)=X(I)
         NIA=II
C    COMPUTE NEW X VALUES
         DO 7 I=1,N
7        X(I)=A*X(I)+(1.-A)*FUN(I,X)
C    CHECK IF STEP LENGTH IS LESS THAN TOL
         STEP=0.
         DO 8 I=1,N
8        STEP=STEP+(X(I)-XX(I))**2
         STEP=SQRT(STEP)
         IF(STEP.LT.TOL) RETURN
5        CONTINUE
         END

C    PROGRAM TO DEMONSTRATE USE OF SUBROUTINE <SUBST>
         PROGRAM TEST
         EXTERNAL FUN
         DIMENSION X(2),XX(2)
         X(1)=1.
         X(2)=2.
         N=2
         NI=20
         TOL=1.E-5
         A=.3
         CALL SUBST(N,NI,NIA,X,XX,FUN,TOL,STEP,A)
         WRITE(*,100) NIA,STEP
100      FORMAT(1X,'AFTER',I5,'ITERATIONS, STEP LENGTH WAS',E12.6)
         WRITE(*,101) X
101      FORMAT(1X,'CURRENT POSITION IS',2E12.6)
         END
C    WE NOW DEFINE FUN(J,X)
         FUNCTION FUN(J,X)
         DIMENSION X(2)
         IF(J.EQ.2) GO TO 5
         FUN=((.5+.0625*X(2)**1.75)/.75)**(4./7.)
         RETURN
5        FUN=((1.-.75*X(1)**1.75)/.0275)**(4./7.)-X(1)
         END


AFTER    10 ITERATIONS, STEP LENGTH WAS .393210E-05
CURRENT POSITION IS .101747E+01 .232246E+01
```

A variant of the successive substitution method, termed *relaxation*, can be effective for large systems. Suppose that we are dealing with a fairly extensive piping network and that pressures or system inflow rates are specified at certain (boundary) nodes. We would like to determine the pressures at all nodes.

We can start by guessing pressure values at those nodes for which pressure is not specified. Corresponding to these pressure values, tentative flow rates in all pipe sections can be calculated by means of equations like those in (3.2). Unless the pressure guesses happen to be correct, the cumulative flow rate into those nodes for which the pressures were guessed will not be correct (i.e., will not vanish, for an interior node, or will not be as specified, for a boundary node). The idea now is to proceed through the network, node by node, adjusting the pressure at each node in turn so as to make the net flow rate into that node correct. (Of course, pressures are kept fixed where specified.) After all node pressures have been "relaxed" in this manner, the entire process is repeated until adequate convergence has been attained. It is sometimes useful to "over-correct" the pressure values at each iteration step; this process is termed *over-relaxation*. The relaxation idea is similar to that used in Section 2.6 in connection with the solution of linear equations. We will encounter the same idea in Chapter 8 when we consider partial differential equations of potential type.

3.4 NEWTON'S METHOD

Consider again the problem of finding x and y such that

$$f(x, y) = 0 , \qquad g(x, y) = 0 \qquad (3.12)$$

and suppose that an initial guess (x_1, y_1) is available. Then f and g have the values $f(x_1, y_1)$ and $g(x_1, y_1)$, which normally will differ from zero. We want to alter x from x_1 to x_2, and y from y_1 to y_2, to bring f and g to zero. Recall that for the case of a function $\phi(x)$ depending on x alone, the change in ϕ corresponding to a change in x from x_1 to x_2 was given to a first approximation by $\phi'(x_1) \cdot (x_2 - x_1)$ (cf. Sec. 1.4). Our present function f depends on both x and y, so that the rates of change with respect to each of x and y must be considered; consequently, the approximate change in f is now given by

$$\left(\frac{\partial f}{\partial x}\right) (x_2 - x_1) + \left(\frac{\partial f}{\partial y}\right) (y_2 - y_1) ,$$

where the two partial derivatives are evaluated at (x_1, y_1). (A partial derivative of the form $\partial f/\partial x$, say, simply means a derivative of f with respect to x, where y is held fixed. A review of partial derivatives is given in Appendix D.)

Since we want to change f and, similarly, g by enough to bring their values from $f(x_1, y_1)$ and $g(x_1, y_1)$ to zero, the approximate equations determining x_2 and y_2 are given by

$$f_x \cdot (x_2 - x_1) + f_y \cdot (y_2 - y_1) = -f(x_1, y_1) ,$$

$$g_x \cdot (x_2 - x_1) + g_y \cdot (y_2 - y_1) = -g(x_1, y_1) , \qquad (3.13)$$

where we follow the common practice of denoting a partial derivative by a subscript (thus g_y means $\partial g/\partial y$). The four partial derivatives in Equations (3.13) are evaluated at the point (x_1, y_1). In matrix form Equations (3.13) can be written

$$\begin{bmatrix} f_x & f_y \\ g_x & g_y \end{bmatrix} \cdot \begin{bmatrix} x_2 - x_1 \\ y_2 - y_1 \end{bmatrix} = - \begin{bmatrix} f(x_1, y_1) \\ g(x_1, y_1) \end{bmatrix} . \qquad (3.14)$$

The coefficient matrix, involving the partial derivatives of f and g, is termed the *Jacobian*; it is often denoted by J.

For the pipe flow problem of Section 3.1, in which f and g are defined by Equations (3.6), the value of J is easily found:

$$J = \begin{bmatrix} 1.3125x_1^{.75} & -.10938y_1^{.75} \\ 1.3125x_1^{.75} + .048125(x_1 + y_1)^{.75} & .048125(x_1 + y_1)^{.75} \end{bmatrix}.$$

Starting again with $x_1 = 1$, $y_1 = 2$ and using Equations (3.13), we obtain a first iteration of $x_2 = 1.0173$, $y_2 = 2.3399$. One more step, with (x_1, y_1) replaced by (x_2, y_2), and (x_2, y_2) replaced by (x_3, y_3) (note that J is now evaluated at (x_2, y_2)) gives $x_3 = 1.0175$, $y_3 = 2.3225$, which is accurate to five significant figures.

The extension to more variables is immediate. Thus the iteration process to solve $f(x, y, z) = 0$, $g(x, y, z) = 0$, and $h(x, y, z) = 0$ becomes

$$\begin{bmatrix} f_x & f_y & f_z \\ g_x & g_y & g_z \\ h_x & h_y & h_z \end{bmatrix} \cdot \begin{bmatrix} x_{j+1} - x_j \\ y_{j+1} - y_j \\ z_{j+1} - z_j \end{bmatrix} = - \begin{bmatrix} f \\ g \\ h \end{bmatrix}, \qquad (3.15)$$

where the functions and their derivatives are evaluated at (x_j, y_j, z_j). Because the mathematical structure of Newton's method is essentially the same as in the single-variable case, we can expect the ultimate convergence rate here also to be quadratic. Also, as in the single-variable case, it may be advisable to truncate unreasonably large iteration steps and, in cases in which it is cumbersome to compute exact derivatives, to approximate them by finite differences (e.g., $f_x \cong [f(x + \delta x, y, \ldots) - f(x, y, \ldots)]/\delta x$). It is also possible to use the same Jacobian matrix for several iteration steps, updating it only occasionally.

3.5 SYSTEM-DETERMINED RESIDUES

Return now to the pipe network problem of Figure 3.1, with the governing Equations (3.2). However, let us denote the flow in branch 3 by q_3 rather than immediately use the fact that it must equal $q_1 + q_2$. Then instead of Equations (3.2) we have

$$p_1 - p_0 = c_1 q_1^{7/4},$$
$$p_2 - p_0 = c_2 q_2^{7/4},$$
$$p_0 - p_3 = c_3 q_3^{7/4} \qquad (3.16)$$

together with, of course, the condition $q_1 + q_2 = q_3$. Again, p_1, p_2, p_3 and c_1, c_2, c_3 are given, and we want to determine the other variables. Previously, we formally reduced the system to a pair of coupled equations in two unknowns and proceeded from there.

A simple idea leads to an alternative procedure. We observe that if we were to make a guess for p_0, Equations (3.16) would give us q_1, q_2, q_3 at once, and our solution would be incorrect only to the extent that

$$\phi = q_3 - (q_1 + q_2) \tag{3.17}$$

were nonzero. This suggests that we think of ϕ as being an error quantity, or *residue*, whose value depends on the choice of p_0 and which is most easily determined by using the numerical value for p_0 in each of Equations (3.16), in turn, and then calculating ϕ by Equation (3.17).

We could certainly write, in this case,

$$\phi = \left(\frac{p_0 - p_3}{c_3}\right)^{4/7} - \left[\left(\frac{p_2 - p_0}{c_2}\right)^{4/7} + \left(\frac{p_1 - p_0}{c_1}\right)^{4/7}\right], \tag{3.18}$$

which is an explicit function of p_0, but in so doing we lose the simplicity of the individual equations in (3.16); also, as previously pointed out, this kind of substitution can be very laborious in large systems. Thus, we discard Equation (3.18) and use Equations (3.16) directly to determine q_1, q_2, and q_3 corresponding to any guessed value of p_0; the residue ϕ is then obtained from Equation (3.17).

We think of ϕ as a function of p_0, writing $\phi = \phi(p_0)$, and iteratively improve p_0 by some such scheme as Newton's method. Generally, we want to avoid formal differentiations, so let us approximate $\phi'(p_0)$ by

$$\phi'(p_0) \cong \frac{\phi(p_0 + \delta p_0) - \phi(p_0)}{\delta p_0}, \tag{3.19}$$

where δp_0 denotes a small change in p_0. What this means is that we slightly alter p_0, compute the corresponding new residue $\phi(p_0 + \delta p_0)$ in exactly the same way as that in which we found $\phi(p_0)$, and then use Equation (3.19) to approximate ϕ'. Newton's method would then lead to the revised value $\bar{p}_0$:

$$\bar{p}_0 = p_0 - \frac{\phi(p_0)}{\phi'(p_0)}.$$

Let us try it (and then we will consider a more complicated example). For $p_1 = 2$, $p_2 = 1.5$, $p_3 = 1$ as before, let us guess $p_0 = 1.2$. Then with $c_1 = .75$, $c_2 = .0625$, $c_3 = .0275$ (also as before), Equations (3.16) give $q_1 = 1.03757$, $q_2 = 2.45065$, $q_3 = 3.10741$, so the residue ϕ equals $q_3 - (q_1 + q_2) = -.38081$. Now alter p_0 to $1.2 + .01$ to obtain, similarly, $\phi = -.238502$, so $\phi' \cong (-.238502 + .38081)/.01 = 14.231$. Thus, the new value for p_0 would be

$$\bar{p}_0 = 1.2 - \frac{-.38081}{14.231} = 1.22676,$$

and we can continue the iteration; one more step will turn out to be adequate.

The general idea is simple but very powerful. As we proceed through some physical system and encounter a quantity whose value is not known, we make a guess for that quantity; thus, we eventually accumulate a number of quantities (call them x, y, z, ...) whose values had to be guessed. But since these guesses are generally in error, we will also find contradictions arising in the system, so we will also accumulate an equal number of error or residue quantities (call them ϕ,

ψ, ω, ...). We think of ϕ, ψ, ω, ... , as being functions of $(x, y, z, ...)$. Our purpose is then to adjust the values of x, y, ... , so as to make these residues vanish. When using Newton's method, we alter x by a small amount δx, leaving y, z, ... unaltered, and determine the alterations in the residues, say $\delta\phi$, $\delta\psi$, ... Then $\delta\phi/\delta x$, $\delta\psi/\delta x$, ... , are approximations to ϕ_x, ψ_x, ... , for use in the Jacobian that arises in Newton's method.

The essential point is that we never use formal algebraic substitution (as in Eq. (3.18), for example); we use only the comparatively straightforward equations governing the various system elements.

In a more complicated piping network, such as that representing the main gas or water distribution system for a city, pressures and withdrawal rates would be specified at some points, and the problem would be to determine the remaining pressures and flow rates. Again, we would begin at some convenient point, using experience to provide good starting guesses (either flow rates or pressures) where required and collecting residues as we work our way through the system. Each guessed quantity, in turn, would be altered slightly, and the corresponding changes in the various residues computed to obtain approximations to the partial derivatives. The requisite alterations in the guesses could then be determined by Newton's method. In a complex system (perhaps several hundred piping sections in the main distribution network), it would clearly not be feasible to treat *all* pressures and flow rates as basic unknown variables, nor would algebraic substitution [as in Eq. (3.18)] be practicable; the effectiveness of the present alternative method then becomes clear. See, however, the warning at the end of this section.

Similar situations arise in electrical networks where ohmic heating may cause resistance values to change. In the circuit of Figure 3.3, let the resistance R_j in the jth branch be given by

$$R_j = R_{j0}(1 + \alpha I_j^2) , \qquad (3.20)$$

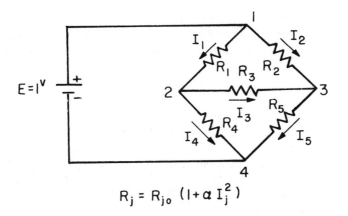

$$R_j = R_{jo}(1 + \alpha I_j^2)$$

Figure 3.3 Nonlinear circuit.

where α and R_{j0} are given constants and I_j is the current in the jth branch. Here electric potential corresponds to pressure in a piping system. If $E = 1$ volt, we take the potential of node 4 as having a reference value of 0, so that the potential of node 1 is $E_1 = 1$. Starting with node 1, let us make a guess for E_2; this then determines I_1 as the solution of

$$E_1 - E_2 = I_1 R_{10}(1 + \alpha I_1^2) .$$

But we also know $E_4 = 0$, so we can use

$$E_2 - E_4 = I_4 R_{40}(1 + \alpha I_4^2)$$

to calculate I_4. We now have numerical values for I_1 and I_4. But the net rate of current flow into node 2 is zero, so from the I_1 and I_4 values we can calculate $I_3 = I_1 - I_4$. Next, we can use

$$E_2 - E_3 = I_3 R_{30}(1 + \alpha I_3^2)$$

to obtain a numerical value for E_3, and in conjunction with E_1 and E_4 this permits us to calculate I_2 and I_5. But we will find that $I_5 - (I_2 + I_3)$, calculated in this way, does not vanish, and this gives us the residue, which we try to reduce to zero by appropriately altering the single guessed quantity E_2. Again, the idea is to avoid algebraic substitutions; everything is done numerically. Problem 3.4 provides a numerical example for the circuit of Figure 3.3.

For a final example, consider the simplified jet engine block diagram shown in Figure 3.4. After pressure recovery in the inlet to a jet engine, a rotating compressor raises the pressure and temperature of the incoming air; fuel is mixed into the airstream and burned in the burner. The hot gas exits at a high velocity through the exit nozzle, thereby providing thrust. Prior to exit, the gas stream drives a turbine, which (through gearing) provides power to the compressor. Air is bled off from the compressor for use in turbine cooling and also for other aircraft purposes; some power is also extracted from the gearing for external use. This diagram is for a single spool engine; a second turbine, driving its own compressor, may be arranged concentrically on the same shaft. In determining the operating conditions (for a given altitude, flight Mach number), there are many equations that must be satisfied. These relate the output of one unit to the input of another

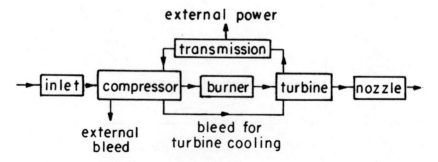

Figure 3.4 Jet engine block diagram (single stage).

and also involve the internal characteristics of each unit. For example, the pressure rise in the compressor depends on the inlet conditions and on the rotational speed in a manner described by sets of empirical graphs (which must, of course, be appropriately stored in the computer). Because high pressures and temperatures are involved, real, rather than ideal, gas properties must be used throughout. We start at the inlet (for given flight conditions) and guess the flow rate into the compressor. With a guess for the compressor rotation speed, we can then use known compressor data to determine compressor exit conditions, which in turn provide the inlet data for the next unit. We proceed in this way through the system, making new guesses as required. Certain discrepancies will arise, because of incorrect guesses (for example, we will probably find that the turbine work output, minus the transmission losses and external power take-offs, will not match the required compressor work input), and we simply note these discrepancies for future use as residues. There is no need to go into further detail, but it is usually possible to reduce the solution of a set of some hundreds of nonlinear equations to a procedure in which not more than about four or five quantities have to be guessed. Each of these quantities is then, in turn, altered slightly and the calculation repeated to obtain numerical approximations for the various partial derivatives, so that Newton's method (with protection against overly large step sizes) can be used for the next iteration step. For a problem of this size, great care must be taken with data management; however, the calculation procedure itself, for each new set of operating conditions, is very fast.

However, a warning is in order. There are many physical systems in which the procedure of this section has proven very effective, but there are also cases in which computational instability may be encountered. One example can be described by reference to Problem 2.18 and Figure 2.4. Suppose we choose $T_{1,1}$, $T_{1,2}$, and $T_{1,3}$ to be the unknowns, guess initial values for them, and then, proceeding to the right, column by column, use the difference equation of Problem 2.18 to compute all other $T_{i,j}$ in terms of these three unknowns. If the reader does this it will turn out that rather large numbers are encountered. Thus we would expect that the present method, applied to a large problem of this kind, could become inaccurate and even unstable.

Sometimes, such difficulties can be resolved by choosing a different set of quantities for use as unknowns. In other cases, it is better to use a different method entirely, such as the relaxation method of Section 3.3. Experience suggests that if there is not much cross-connection between elements of the system (as in the jet engine example), then the method of this section can be powerful; if there is a lot of cross-connection (as in an extended version of Figure 2.4), then relaxation may be the method of choice.

3.6 USE OF A PARAMETER

With reference to Figure 3.5, suppose we wanted to find the shortest distance between the parabola $y = x^2$ and the straight line $y = -1 + \frac{1}{2}x$. Let the shortest

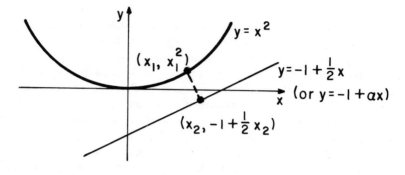

Figure 3.5 Parameter idea.

line be that which joins (x_1, x_1^2) on the first curve to $(x_2, -1 + \frac{1}{2}x_2)$ on the second. Then we want to find x_1 and x_2 to minimize

$$D = (x_1 - x_2)^2 + (x_1^2 - [-1 + \frac{1}{2}x_2])^2 .$$

suppose we knew the correct value of x_1. Then D would be a function of x_2 only, and the condition that D be a minimum for a particular value of x_2 is that the derivative of D with respect to x_2 vanish at that point. Since we could similarly hold x_2 fixed and vary x_1, it follows that a necessary condition that a minimum be achieved for a certain choice of (x_1, x_2) values is that $\partial D/\partial x_1 = 0$, and $\partial D/\partial x_2 = 0$, at that point. Thus, our problem is to find x_1 and x_2 so that

$$\frac{\partial D}{\partial x_1} = 2(x_1 - x_2) + 2(x_1^2 - [-1 + \frac{1}{2}x_2]) \cdot (2x_1) = 0 ,$$

$$\frac{\partial D}{\partial x_2} = -2(x_1 - x_2) + 2(x_1^2 - [-1 + \frac{1}{2}x_2]) \cdot (-\frac{1}{2}) = 0 . \qquad \textbf{(3.21)}$$

We could use Newton's method to solve this pair of equations, and it is not too hard to find a reasonable starting guess. However, in a complicated problem involving several nonlinear equations, it could be troublesome to come up with a good starting guess—and without a good start, any iterative method could lead us astray. A device that is often useful is to introduce a parameter into the problem. In the present example let us alter the problem by changing the line to $y = -1 + \alpha x$, where α is a parameter. We are eventually interested in the case $\alpha = \frac{1}{2}$, but let us start with $\alpha = 0$, for which case we know that the solution is given by $x_1 = 0$, $x_2 = 0$. Now increase α slighlty, say to $\alpha = .05$. Then in the problem that corresponds to Equations (3.21) (which now involve α, with α set equal to .05), the choice $x_1 = 0$, $x_2 = 0$ should provide an excellent starting guess for Newton's method, and the solutions for x_1 and x_2 corresponding to $\alpha = .05$ could be determined rapidly. We then use these new values of x_1 and x_2 as starting values for a new problem in which $\alpha = .1$, and so on; eventually we reach $\alpha = \frac{1}{2}$.

Although the parabola–straight line problem is a rather simple one, it should make the basic idea clear. One question that arises in a more complicated problem is, How does one introduce a suitable parameter α? The answer is that this is usually a matter involving some ingenuity. However, a number of standard approaches have been suggested. For the problem $f(x, y) = 0$, $g(x, y) = 0$, one method is to start with guesses x_0 and y_0 and then form

$$F(x, y, \alpha) = f(x, y) - \alpha f(x_0, y_0) ,$$

$$G(x, y, \alpha) = g(x, y) - \alpha g(x_0, y_0) .$$

Then for $\alpha = 1$, we know a solution to $F = 0$, $G = 0$; it is $x = x_0$, $y = y_0$. We start at this point and solve a sequence of problems $F = 0$, $G = 0$, in which α is gradually reduced; when we reach $\alpha = 0$, the problem $F = 0$, $G = 0$ has become equivalent to $f = 0$, $g = 0$. Somewhat more scope would be provided by the choice $F = (x - x_0)\alpha + (1 - \alpha)f$, $G = (y - y_0)\alpha + (1 - \alpha)g$. There are many other possibilities. Of course, nothing is guaranteed—any one of these choices for F and G could turn out to lead us down the garden path.

ANNOTATED BIBLIOGRAPHY

F. S. Acton, 1970, *Numerical Methods that Work*, Harper & Row, New York, 539p.

Chapter 14 presents an informal but provocative approach to nonlinear equation solving. On page 380, you will encounter the timid curve crawler.

B. Carnahan, H. A. Luther, and J. O. Wilkes, 1969, *Applied Numerical Methods*, John Wiley, New York, 604p.

A detailed discussion of the nonlinear pipe flow problem, which includes a suitable mechanism for describing which nodes are connected by pipes, is given on page 310. A second example dealing with chemical reactions will be found on page 321. Both flow diagrams and computer programs (using Newton's method) are provided.

P. Rabinowitz, ed., 1970, *Numerical Methods for Nonlinear Algebraic Equations*, Gordon and Breach, London, 199p.

This book contains contributions from a number of authors and describes experiences with and perspectives on nonlinear equations. Many of the papers include useful bibliographies.

PROBLEMS

3.1 Write a computer subroutine that uses the secant plane method of Section 3.2 to solve a set of two nonlinear equations and, as a check, apply it to the pipe network problem of Section 3.1. Make use of the modification suggested in the discussion of Equations (3.6).

3.2 Write a subroutine that solves n nonlinear equations by Newton's method. Compute the partial derivatives by an approximation method in which each variable, in turn, is incremented by an amount d (an input from the calling program). Include protection against overly large iteration steps.

3.3 Can the bisection method of Section 1.2 be extended to apply to the case $f(x, y) = 0$, $g(x, y) = 0$? (*Hint*: One possiblity is to consider values of g as one follows along (x, y) path defined by $f(x, y) = 0$.) Discuss.

3.4 Use any method to solve the network problem of Figure 3.3, given that $R_{10} = 1$, $R_{20} = R_{40} = \frac{1}{2}$, $R_{30} = .7$, $R_{50} = 3$, $\alpha = 1$, in Equation (3.20). Verify your solution.

3.5 A student suggests that if the potentials of nodes 2 and 3 in Figure 3.3 are taken as unknowns, then, since $E_4 = 0$ (as a reference potential) and $E_1 = 1$, all of the currents could be determined, in principle, as functions of E_2 and E_3. However, there are three conditions these currents must satisfy: (a) zero net current at node 2, (b) zero net current at node 3, and (c) the net current flowing out of node 4 and into the battery must equal the current flowing out of the battery into node 1. Since we thus have three conditions but only two variables, the problem seems overdetermined. How could you resolve his difficulty?

3.6 Suppose that a number n of points, say $(x_1, y_1, z_1), \ldots, (x_n, y_n, z_n)$ are obtained by a measurement process, and suppose that we know that, apart from measurement errors, these points should lie more or less on some plane in space. We want to find the plane that most closely fits this set of points, in the sense that the sum of the squares of the normal distances of the points from the plane should be a minimum.

From the origin draw a line perpendicular to the solution plane; let this line make an angle θ with the z-axis, and let the projection of the line onto the (x, y) plane make an angle ψ with the x-axis. Moreover, let the length of this normal line be r. Then we want to find r, θ, ψ. Now for any one of the points, say (x_i, y_i, z_i), the distance d_i of that point from the plane would be given by

$$d_i^2 = (x_i \sin \theta \cos \psi + y_i \sin \theta \sin \psi + z_i \cos \theta - r)^2 .$$

Thus, the sum of these squared distances can be formed, and by setting the partial derivatives with respect to each of r, θ, ψ equal to zero, we get conditions that determine the optimal plane. Devise a technique and write a program that solves this problem.

3.7 Find all five roots of

$$(1 + i)z^5 - 2z^3 + iz^2 - 1 = 0 .$$

(*Hint*: Write $z = x + iy$, where x and y are real, and separate real and imaginary parts to obtain a pair of real equations in x and y.

3.8 In polar coordinates (r, θ), the equation of an elliptical orbit, corresponding to motion in the gravitational field of a mass placed at the origin, is given by

$$\frac{1}{r} = A(1 + \epsilon \cos(\theta - \theta_0)),$$

where the eccentricity ϵ satisfies the condition $0 < \epsilon < 1$ and the constant θ_0 represents the value of θ at which r is least. The quantity A is a constant of the motion. For two satellites moving around the earth, describe an efficient technique for determining the shortest distance between their orbits. Take the motion of the two satellites as being in the same plane. Can the parametric idea of Section 3.6 be effective here?

3.9 In applying various methods to the problem described by Equations (3.6), we usually started with the rather good guess $x_1 = 1$, $y_1 = 2$. Try quite different starting values and see what happens. (*Note*: You may want to replace $q^{7/4}$ by $q|q|^{3/4}$ in the original problem of Sec. 3.1 in order to make the final equations valid for either flow direction.)

3.10 In the gas distribution network of Figure 3.6, the flow through each of the seven pipe sections is related to the pressure difference across its ends by

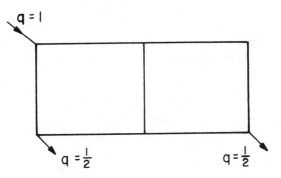

Figure 3.6 Distribution network for Problem 3.10.

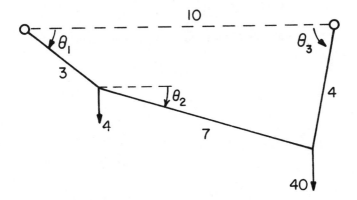

Figure 3.7 Support system for Problem 3.11.

equations of the form considered in Section 3.1. Nondimensionalizing as in Section 3.1, let input and output flow rates be as shown in Figure 3.6; take all $c_i = 1$. Find the pressures at all nodes (relative to the pressure at the input node). Try the methods of Sections 3.3 (relaxation) and 3.5.

3.11 A wire of length 14 units is attached to two support points placed on the same horizontal line at a distance of 10 units apart. Two weights, of 4 units and 40 units, are suspended from the wire at distances of 3 units and 4 units from the two ends, as shown in Figure 3.7. Use an efficient method to find the angles θ_1, θ_2, θ_3 indicated in the figure, as well as all wire tensions. Repeat for the case in which an added wieght of 14 units is suspended halfway between the original weights.

3.12 In electrical tomography it is desired to determine internal resistivities by surface measurements. A suggested model situation is depicted in Figure 3.8. Here seven resistors, of unknown values $R_1, R_2, \ldots, R_7$, are connected in a loop. The resistance across each pair of adjacent nodes is measured, so as to give seven values $\rho_1, \rho_2, \ldots, \rho_7$. Using the known ρ_i values in the seven equations

$$\rho_i = \frac{R_i(R - R_i)}{R},$$

where $R = R_1 + R_2 + \cdots + R_7$, we must find the R_i. Write a program that will do this for the case of N resistors in a loop. Provide appropriate test cases.

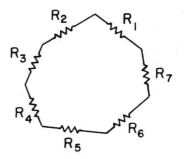

Figure 3.8 Model for electrical tomography.

4

RANDOM NUMBERS IN NUMERICAL ANALYSIS

One often encounters practical problems in which some degree of randomness is present. Examples include radioactive decay, turbulence, signal processing in the presence of noise, games of chance, traffic control, inventory management, and Brownian motion. In computer simulation of situations such as these, it is useful to be able to generate random numbers as needed; Section 4.1 describes methods for doing this. Section 4.2 discusses the use of random numbers to obtain estimates for areas, volumes, and moments of inertia for regions of complicated shape. Subsequent sections consider the accuracy of such estimates and methods for generating random numbers with Poisson or normal distributions. A number of applications, in simulation and in numerical analysis, are given as problems at the end of the Chapter.

4.1 COMPUTER GENERATION OF RANDOM NUMBERS

Since the output of a computer is determined in a predictable and repeatable way by its set of instructions, the use of a computer to generate random numbers seems at first sight to be paradoxical. Remarkably enough, however, it turns out that there are a number of iterative processes that generate sequences of numbers that an observer, unaware of their origin, would consider to be quite random.

For example, consider the set of 100 computer-generated random numbers between 0 and 1 given in Table 4.1. There does not seem to be any feature of this set that would distinguish it from a similar one obtained, say, by randomly drawing numbers from a barrel.

Table 4.1.   One-hundred computer-generated, random numbers between 0 and 1

.2534	.9293	.7007	.9032	.6696
.6523	.3038	.6562	.8455	.3145
.9067	.6342	.6742	.2249	.2233
.1860	.5872	.4205	.9227	.2632
.2714	.8251	.2937	.8733	.2036
.1377	.1949	.4201	.1881	.5145
.9776	.4927	.0860	.1068	.0681
.6671	.9232	.5154	.8983	.8805
.5654	.2274	.8716	.9442	.1459
.0939	.0504	.7542	.4360	.9153
.2552	.7643	.3436	.1926	.2605
.2834	.7285	.1106	.7756	.1588
.7211	.8133	.8988	.4099	.7336
.0352	.7314	.3725	.3486	.1501
.8393	.9180	.2755	.7009	.2499
.3091	.5600	.4720	.6714	.2548
.2353	.1671	.6227	.7268	.6019
.7557	.1604	.1007	.3243	.5487
.8915	.1832	.5314	.0685	.9584
.6894	.1078	.7736	.6843	.7308

The method used to generate the numbers in Table 4.1 can be illustrated by the following simple example. Choose some fixed multiplier, say $m = 33$. Next, start with some integer, say 17, and multiply it by m to give 561. Keep only the last two digits, 61, and again multiply by m to give 2013. Again keep only the last two digits, 13, and again multiply by m. Continuing in this way, we generate the sequence 17, 61, 13, 29, 57, 81, 73, 09, 97, 01, 33, 89, 37, 21, ... Now place a decimal point ahead of each, and we end up with a set of numbers in the interval (0, 1) that possesses a certain randomness. Of course, this particular sequence has some defects. It would eventually repeat itself, and also the last digit cycles through the sequence 7, 1, 3, 9. However, when the method is extended so as to use more digits, defects such as these become unimportant.

If the nth number in the example is denoted by x_n, then the process is described by

$$x_{x+1} = mx_n \ (\text{mod } 100) \ ,$$

where the term (mod 100) or (modulo 100) means that we subtract as many integral multiples of 100 as we can, still keeping the result positive. More generally, m need not be 33, and we can keep p digits rather than two; this would correspond to $x_{n+1} = mx_n \ (\text{mod } 10^p)$. Since most computers use binary numbers rather than decimal, it is usually convenient to reduce all numbers by subtracting multiples of 2^q rather than 10^p. If q is taken as the word length of the computer, then the unwanted digits are automatically discarded when the computer does integer arithmetic.

Random numbers can be generated in other ways. For example, we could go from an integer x_n to x_{n+1} by squaring x_n and keeping only a certain number of the central digits in the result. Alternatively, addition, rather than multiplication, or a combination of the two, could be used to form x_{n+1} from x_n. However, because of its speed and simplicity, the choice

$$x_{n+1} = mx_n \;(\text{mod } 2^q) \qquad (4.1)$$

is often made, where q is the number of bits in a computer word.

Because the higher-order bits are discarded after each multiplication, the process of Equation (4.1) yields numbers that (other than the lowest-order bits) are unpredictable—except by the generating process itself. Thus, Equation (4.1) can serve as a simple and fast source of pseudo-random numbers. Of course, some precautions are necessary. The multiplier m should be odd in order to avoid the eventual generation of nothing but zeros. Also, the initial number x_1 should be chosen odd if we want to utilize the full word length of the computer. Observe also that since the computer word can hold only a finite number of different numbers, the sequence generated by Equation (4.1) must eventually repeat itself; it is desirable to make the period of the sequence as long as possible. For Equation (4.1) it turns out from number theory (see Knuth, 1969) that if m has the form $8k + 5$ or $8k + 3$, where k is some integer, then Equation (4.1) will generate 2^{q-2} numbers before the sequence starts to repeat itself. The number m should also be reasonably large in order to avoid starting off the sequence (for some choices of x_1) with a set of small numbers.

Unfortunately, ANSI FORTRAN77 does not require a random number generator to be included in its list of intrinsic functions. Because word length is involved, the particular form depends on the computer used. Many computer installations do in fact provide a random number generator as an intrinsic function (often written in machine language), and the reader may wish to use such a function if available.

The following program would be suitable for a computer that has a word length of 48 bits.* The user provides an initial "seed" number ISEED; from then on, the program reseeds itself. About 7×10^{13} numbers will be generated before the sequence starts to repeat itself.

```
C   SUBROUTINE RAND(ISEED,N,RD) FILLS AN ARRAY RD
C   WITH N RANDOM NUMBERS IN INTERVAL (0,1).  AN
C   INITIAL SEED INTEGER ISEED MUST BE PROVIDED;
C   ISEED MUST BE ODD AND POSITIVE.  THE ROUTINE
```

---

*For a microcomputer with a 2-byte integer word length, with numbers stored in two's complement form, one could use IX = 31691, say, with RD(I) = FLOAT(ISEED)/65535. + .5. The entering value of ISEED should be in the range $(-32767, 32767)$. About 16,000 numbers will be generated before the sequence starts to repeat.

```
C   MAY BE RESEEDED AT ANY TIME BY GIVING A NEW
C   ISEED; OTHERWISE, THE ROUTINE RESEEDS ITSELF.
C
        SUBROUTINE RAND(ISEED,N,RD)
        DIMENSION RD(N)
        IX=454806270215 23
        DO 7 I=1,N
        ISEED=ISEED*IX
7       RD(I)=FLOAT(ISEED)/281474976710655.
        END
```

4.2 AREAS, VOLUMES, AND MOMENTS

One application of random numbers is in the approximate calculation of areas, volumes, and moments of inertia. As an initial example, the area of a quarter-circle will be determined; the result can be used to find an estimate for π. Consider a quarter-circle enclosed within a unit square as shown in Figure 4.1a. Suppose that we generate pairs (x, y) of random numbers, where x and y lie anywhere in the range $(0, 1)$. Then the point with coordinates (x, y) can fall anywhere in the unit square, and if we generate a large number N of such pairs, the corresponding points should cover the unit square with more or less uniform density. That fraction of the points within the shaded area should approximate the ratio of the shaded area $(\pi/4)$ to the area (1) of the square, presumably more and more closely as N increases. Of course, the test to see whether a particular random point (x, y) is in the shaded area is simply to test the value of $x^2 + y^2$. In three trials of this kind, using 10,000 points each time, it was found that 7714, 7801, and 7841 points fell in the shaded region; the corresponding approximate values of π are 3.0856, 3.1240, and 3.1364. In a second set of trials using 100,000 points, the numbers were 78,535, 78,782, and 78,675, with $\pi \cong 3.14140$, 3.15128, and 3.14700, respectively.

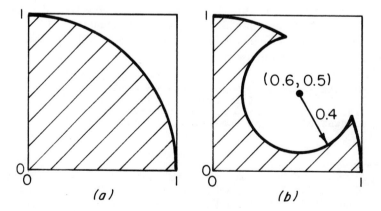

Figure 4.1. Area calculation.

Except in a trial case, one would hardly use this method—sometimes termed, for obvious reasons, a *Monte Carlo method*—to determine the area inside a quarter-circle. However, things become more interesting if we take a bite out of the circle. In Figure 4.1*b*, that part of the original quarter disk that falls inside the circle of radius .4, centered at (.6, .5) has been removed. It is now more cumbersome to determine the shaded area by calculus, but in the Monte Carlo method we need only alter the acceptance rule, so that an (x, y) point is "counted" if it simultaneously satisfies the two conditions $x^2 + y^2 < 1$, $(x - .6)^2 + (y - .5)^2 > .4^2$. The reader is invited to try his fortune. The usefulness of the Monte Carlo method increases still more if we increase the number of dimensions. A typical three-dimensional problem might be to determine how much volume is removed from the interior of a sphere of radius R by passing a straight circular pipe of radius r through the sphere, the centerline of the pipe being a distance a from the center of the sphere. Place the origin of the (x, y, z) coordinates at the center of the sphere, with the centerline of the pipe parallel to the x-axis and lying in the (x, y) plane. Enclose the sphere in a cube of side $2R$ and generate random numbers w in the range $(-R, R)$ [if α is a random number in (0, 1) as generated by RAND, say, then use $w = R(2\alpha - 1)$]. Use these numbers in sets of three to provide points lying randomly in the cube, and count how many of these simultaneously satisfy the conditions $x^2 + y^2 + z^2 < R^2$, $(y - a)^2 + z^2 < r^2$. Of course, we could save ourselves some effort by removing the lower half of the cube from consideration, and still more by using symmetry; so we need consider only the region $x > 0$, $z > 0$, $y > 0$.

The Monte Carlo method can also be used to estimate moments of inertia of areas and volumes. Suppose we want the polar moment of inertia, about the origin, of the shaded region of Figure 4.1*b*. We must consider each element of area dA in the shaded region, multiply dA by the square of its distance from the origin, and combine these products for all such elements of area. Let us generate N random points (x_j, y_j) lying in the unit square. Then any element of area, dA_j, in the shaded region can be expected to contain $(dA_j/1)N$ of the points. For each point accepted as lying in the shaded area, compute $x_j^2 + y_j^2$ and add together all of these contributions. Corresponding to the element dA_j, since there are about $N\,dA_j$ points in that area element, the total contribution arising from these points will be $(N\,dA_j)(x_j^2 + y_j^2)$, at least approximately. Altogether, our sum equals $\Sigma_j N\,dA_j(x_j^2 + y_j^2)$, so we have to divide the grand total by N to get what we really want, which is $\Sigma_j dA_j(x_j^2 + y_j^2)$ or, in the limit, $\int dA(x^2 + y^2)$. See Problem 4.5.

4.3 ACCURACY

A question that occurs to anyone using the Monte Carlo approach of the preceding section is, If an answer with an expected accuracy of, say, 1% is desired, how many points should be used?

Consider again the quarter-circle problem of Figure 4.1*a*. In eight experiments, each using a different set of 10,000 randomly generated points lying in the unit square, the numbers of points inside the quarter-circle were found to be 7831, 7909, 7910, 7880, 7818, 7846, 7892, and 7830. Since the area of the quarter-circle is $\pi/4 = .7853982$ of the area of the square, the expected number of points in the quarter-circle would be $(10,000) \cdot (.7853982) = 7854$. The deviations from this number in the eight experiments were therefore $-23, 55, 56, 26, -36, -8, 38,$ and -24. As a measure of the average deviation, we could take the average of the absolute values of these quantities. For analytical work, however, absolute values are not well-behaved functions (think of a plot of $|x|$ versus x near the origin), and it is preferable to obtain positive quantities by squaring the above deviations (x^2 is a smooth function of x). We then average these squares and take the square root of this average. This root-mean-square quantity is generally denoted by σ and called the *standard deviation*. In the present case we have

$$\sigma = \sqrt{\tfrac{1}{8}\,[(-23)^2 + (55)^2 + \cdots + (-24)^2]} = 36.7 \ . \qquad (4.2)$$

We think of σ as a slightly distorted, but analytically convenient, measure of the average deviation. Its square, σ^2, is termed the *dispersion* or *variance*. (In many cases, we do not know the theoretically expected value; σ and σ^2 could then be calculated on the basis of deviations from the mean value of the experimental results.)

A repetition of the eight experiments, where 10^6 points were used in each experiment, gave (for the numbers of points lying in the shaded region of Fig. 4.1a) 785,931, 784,782, 785,951, 785,502, 785,362, 785,446, 785,273, and 785,204. The expected number would be $(10^6)\, \pi/4 = 785,398$, so the discrepancies are $533, -616, 553, 104, -36, 48, -125,$ and -194. The corresponding value of σ now turns out to be 360.0, so that the average discrepancy from the expected value is substantially larger than in the 10,000-point case. In using the results to calculate $\pi/4$, however, we have to divide by the total number of points, which is now 10^6 instead of 10^4, and it is clear that the relative effect of the discrepancy is much less for the 10^6 case.

Now let us put all of this on a scientific basis. Suppose that, instead of 10^4 or 10^6 trials in each experiment, we use N trials. Then *it will turn out that σ can be expected to be proportional to* $\sqrt{N}$. When we divide by N, to determine the desired area, the error in the result will be proportional to $\sqrt{N}/N$ or $1/\sqrt{N}$. To halve the expected error requires a quadrupling of the number of experiments, and this is an important consideration in using the Monte Carlo method.

We now prove this result. We first define the *expected value* of a quantity to be the average of the quantity over a very large (strictly speaking, approaching an infinite) number of experiments. Thus, for the expected value n of the number of points falling in the shaded region of Figure 4.1a, out of N points altogether in each experiment, we would write $E(n) = Np$, where the sumbol E means expected value and $p = \pi/4$. The reason for this, of course, is that a randomly chosen point could fall with equal probability anywhere in the unit square, and the quarter-circle occupies a portion p of that square, so that, as the number of experiments (each using N points) is increased indefinitely, we would be confident that an overall fraction p of the points would lie in the quarter-circle.

In any one experiment we have N individual trials. With the ith trial, associate a number x_i, defined to be 1 if the random point falls within the quarter-circle, and 0 otherwise. Then the total number of points falling within the circle is $x_1 + x_2 + \cdots + x_N$, and, by definition,

$$\sigma^2 = E\{[x_1 + x_2 + \cdots + x_N - Np]^2\}$$
$$= E\{[(x_1 - p) + (x_2 - p) + \cdots + (x_N - p)]^2\}$$
$$= E\{(x_1 - p)^2 + (x_2 - p)^2 + \cdots + 2(x_1 - p)(x_2 - p) + \cdots\}$$
$$= E\{(x_1 - p)^2\} + E\{(x_2 - p)^2\} + \cdots + E\{2(x_1 - p)(x_2 - p)\} + \cdots .$$

The reason for the validity of the last step here is that the individual trials are independent of one another. But since there is no relation between x_1 and x_2, for example, it follows that $E\{(x_1 - p))(x_2 - p)\} = 0$ (for any possible value of x_1, the average of $x_2 - p$ over many experiments must vanish), so that the cross product terms give no contribution. We then have

$$\sigma^2 = E\{(x_1 - p)^2\} + E\{(x_2 - p)^2\} + \cdots + E\{(x_N - p)^2\} . \qquad \textbf{(4.3)}$$

Consider a typical term:

$$E\{(x_j - p)^2\} = E\{x_j^2 - 2x_j p + p^2\} = E(x_j^2) - 2pE(x_j) + p^2 .$$

But x_j^2 equals 1 or zero whenever x_j does, so $E(x_j^2) = p$, and we get

$$E\{(x_j - p)^2\} = p - 2p(p) + p^2 = p - p^2 .$$

We obtain finally, from Equation (4.3), that $\sigma^2 = Np(1 - p)$, or

$$\sigma = \sqrt{Np(1 - p)} ,$$

which is the desired result, and the derivation is complete.

To summarize, if the probability of an event (such as a point falling in the shaded region of Fig. 4.1) is p, then in N trials the standard deviation should be about $\sqrt{Np(1 - p)}$. This quantity can be thought of as representing the average discrepancy from the expected value Np.

With $p = \pi/4$ and $N = 10,000$, then $\sqrt{Np(1 - p)} = 41.05$, and this corresponds reasonably well with the value $\sigma = 36.7$ obtained in Equation (4.2). With $N = 10^6$, the value of $\sqrt{Np(1 - p)}$ becomes 410.5, which may be compared to the experimental value $\sigma = 360$ reported earlier.

To obtain the expected relative error, we divide the discrepancy quantity $\sqrt{Np(1 - p)}$ by the expected number of "successes" Np to get

$$\frac{\sqrt{Np(1 - p)}}{Np} = \sqrt{\frac{1 - p}{Np}} .$$

For $p = \pi/4$ and $N = 10,000$, the expected fractional error is about $\frac{1}{2}$ of 1%.

We have assumed in this section that the random points are indeed random; whether they are or not depends on the quality of the random number generator. A number of tests for quality are available. A simple one consists in dividing the interval (0, 1) into a number of subintervals and recording the proportion of random numbers falling within each subinterval (cf. Prob. 4.1). Another evaluation is based on the chi-square test (cf. Prob. 4.2). Of course, any use of a random number generator to approximate a known result, as in the previous evaluation of π, is itself a test of quality.

4.4 GENERATION OF OTHER DISTRIBUTIONS

Until now, we have been primarily concerned with the generation and use of random numbers x lying in the interval (0, 1). Since any value of x in this interval is equally likely, we say the x-values are *uniformly distributed* in (0, 1). In fact, the probability that the next x-number will fall within some subinterval (a, b) contained in (0, 1) is just the ratio of the size of this subinterval to the size of the unit interval and, therefore, equals $b - a$. As a special case of this, the probability that the next random number will be found within some interval dx (lying in (0, 1)) is equal to dx. Equivalently, the probability $P(a)$ that x will be less than some chosen number a in (0, 1) is equal to a.

One exception to the use of the unit interval arose in Section 4.2, where we wanted to generate random numbers distributed uniformly in the interval $(-R, R)$. More generally, suppose we want to generate uniformly distributed random numbers y lying in the interval (α, β) where α and β are any chosen real numbers (positive or negative) satisfying $\alpha < \beta$. Then it is clear that we can simply choose x randomly in (0, 1) and define $y = \alpha + (\beta - \alpha) x$.

More generally still, suppose that we want to generate random numbers y such that $P(\alpha)$, the probability that y will be less than α, is some prescribed nondecreasing function of α. For example, we could choose

$$P(\alpha) = 0 \qquad \text{for } \alpha < -2 \,,$$

$$P(\alpha) = \frac{1}{2} \left(1 + \sin \frac{\pi\alpha}{4} \right) \quad \text{for } |\alpha| < 2 \,,$$

$$P(\alpha) = 1 \qquad \text{for } \alpha > 2 \,. \tag{4.4}$$

The function $P(\alpha)$ is plotted in Figure 4.2. Starting again with numbers x distributed uniformly in (0, 1), how should we relate y to x? The idea is remarkably simple: All we have to do is associate x- and y-values to one another by the formula $P(y) = x$. To show that this correspondence between y and x is indeed appropriate, observe first that if ξ is any chosen fixed number between 0 and 1, and if the next random number generated has a value x, then the a priori probability that x will be less than ξ is just equal to ξ. Thus, if α is any chosen number in (0, 1), the probability that $\{x < P(\alpha)\}$ is equal to $P(\alpha)$. Since $x = P(y)$, this statement is

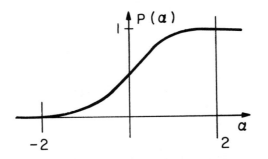

Figure 4.2. Example distribution.

equivalent to the statement that the probability $\{P(y) < P(\alpha)\}$ is equal to $P(\alpha)$. But $P(y)$ is a nondecreasing function of y, so instead of writing $\{P(y) < P(\alpha)\}$ in the last sentence, we could equally well write $\{y < \alpha\}$, and the proof is now complete.

In the example of Equation (4.4), we set

$$x = \frac{1}{2}\left(1 + \sin\frac{\pi y}{4}\right) \quad \text{or} \quad y = \frac{4}{\pi}\arcsin(2x - 1) ,$$

where $-\pi/2 < \arcsin(2x - 1) < \pi/2$. The same method works in the general case for any given $P(\alpha)$ (provided, of course, that $P(\alpha)$ is a feasible probability function; i.e., $P(\alpha)$ does not decrease as α increases, and $P(\alpha) \to 1$ as $\alpha \to \infty$).

Thus, to generate random y-numbers satisfying the condition that $P(\alpha)$, the probability of the next y-number being less than α, is some given function of α, we generate an x-number in $(0, 1)$ and find y to satisfy $P(y) = x$. If the interval (a, b) is part of the possible range of y-values, then the probability that the next y-value will fall within this interval is just $P(b) - P(a)$. If $b \to a$, then $P(b) - P(a) \to P'(a) \cdot (b - a)$, by the definition of a derivative. This says that the probability that y be found in the interval $(y, y + dy)$ is equal to $P'(y) \cdot dy$. The function $P'(y)$ is often denoted by $p(y)$ and termed the *probability density function*.

In generating random numbers to correspond to a desired distribution, there are many tricks of the trade. To illustrate, suppose we want $P(\alpha) = 0$ for $\alpha < 0$, $P(\alpha) = \alpha^2$ for $0 < \alpha < 1$, $P(\alpha) = 1$ for $\alpha > 1$. Then the above method would involve the use of $x = y^2$ or $y = \sqrt{x}$. Problem 4.12 asks the reader to show that, alternatively, one could generate x-numbers in pairs—say, first, x_1 and, second, x_2. If $x_2 < x_1$, then x_1 is accepted as the y-number; otherwise one starts all over again with a new pair. The numbers generated in this way turn out to have the desired distribution. As a second example, suppose one wants to choose directions randomly in the plane, and, perhaps as part of a larger problem, compute the corresponding sines and cosines. Then one method is to choose θ randomly in the interval $(0, 2\pi)$ and to calculate the sine and cosine of the resulting angle. Another method is to choose points (x, y) randomly in the region $-1 < x < 1$, $-1 < y < 1$ and to discard any choice for which $x^2 + y^2 > 1$. The resulting points should be uniformly distributed within the unit circle. These points, when linked to the origin, determine a direction, and the cosine and sine quantities are simply x/r, y/r, where $r = \sqrt{x^2 + y^2}$. In the same way we could choose a random direction in space by generating triplets (x, y, z) in the region $-1 < x < 1$, $-1 < y < 1$, $-1 < z < 1$ and discarding any points for which $x^2 + y^2 + z^2 > 1$. A neutron diffusion problem, in which this method could be used, is described in Problem 4.10.

4.5 POISSON DISTRIBUTIONS

Suppose an event occurs n times per second on the average but otherwise occurs randomly. Radioactive decay provides one example of this kind of situation; here the "event" is the emission of a particle. Another example would be that of incoming telephone calls to a switchboard. Still another could be that of customer arrival at a counter. (Although we use a second as the basic time unit here, the time unit could as easily be an hour, a week, or a year.)

We now ask, What is the probability P_k that exactly k events will occur in a time interval of t seconds? We will prove that this probability is given by the *Poisson distribution*, defined by

$$P_k = \frac{(nt)^k}{k!} e^{-nt} \qquad (4.5)$$

(where $0! = 1$). To give some feeling for the numbers involved, suppose that, on the average, four events happen per second, so $n = 4$. Let $t = 2.5$ seconds. Then Table 4.2 gives the values of P_k for various choices of k, as provided by Equation (4.5). If we were to repeat an experiment many times, counting each time how many events occurred in 2.5 seconds, we should get an average of 10 events—after all, this is how the average rate of four events per second is determined. Note, however, that the probability of finding *exactly* 10 events in any one 2.5-second interval is not large: It is only about 1 in 8.

The derivation of Equation (4.5) is straightforward. Divide the time interval t into a large number, say N, of subintervals. On the average there will be nt events in the interval t, and (if the subintervals are small enough that the probability of

Table 4.2. **Poisson distribution**

P_0	.000 045	P_6	.063 055	P_{12}	.094 780
P_1	.000 454	P_7	.090 079	P_{13}	.072 908
P_2	.002 270	P_8	.112 600	P_{14}	.052 077
P_3	.007 567	P_9	.125 110	P_{15}	.034 718
P_4	.018 917	P_{10}	.125 110	P_{16}	.021 699
P_5	.037 833	P_{11}	.113 736	P_{17}	.012 764

more than one event occurring within any one subinterval is negligible) we can expect on the average that nt of the N subintervals will have events. Thus, the probability that any one subinterval, of length $\delta t = t/N$, will have an event is just $nt/N = n\delta t$. *Given a small time interval dt, the probability that an event will occur within that time interval is n dt.*

Now, what is the probability that *no* events occur within the time interval t? Again, divide t into a large number N of subintervals δt. No event can occur in the first subinterval, and the probability of this is $p_1 = 1 - n\delta t$. Next, no event can occur in the second subinterval, so we have to multiply p_1 by $1 - n\delta t$ to get the probability $(1 - n\delta t)^2$ that no event occurs within the first two subintervals. Continuing in this way, we find

$$P_0 = (1 - n\delta t)^N$$

$$= \left(1 - n\frac{t}{N}\right)^N$$

$$= \left[\left(1 - \frac{1}{W}\right)^W\right]^{nt},$$

where $W = N/nt$. In the limit, as $N \to \infty$ (therefore $W \to \infty$), we use the fact that the quantity in brackets $\to e^{-1}$ (this is one of the ways of defining e, the base of natural logarithms) to get

$$P_0 = e^{-nt}.$$

This, then, is the probability that no events occur in time t.

Next, what is P_1, the probability that exactly *one* event occurs in time t? Again use N subintervals. If one event occurs, it could occur in any one of these N intervals, in each case with probability $n\delta t$. We must now multiply the resulting overall probability, $(N)(n\delta t)$, by the probability that no events occur in any of the other intervals, and as $N \to \infty$ this will clearly be e^{-nt} again. Thus,

$$P_1 = \lim_{N \to \infty} (Nn\delta t)\, e^{-nt}$$

$$= \lim_{N \to \infty} \left(N\frac{nt}{N}\right) e^{-nt}$$

$$= nt\, e^{-nt}.$$

The next case is that in which two events occur. With N subintervals there are $N(N-1)/2!$ different ways of choosing two subintervals; multiplying by the probability that an event will occur in both (i.e., by $(n\delta t)^2$), and again by the probability that no event occurs in the other subintervals (which has a limiting value of e^{-nt}), we get

$$P_2 = \frac{(nt)^2}{2!} e^{-nt}$$

as $N \to \infty$. The proof of Equation (4.5) for other values of k proceeds analogously.

If we want to use a Poisson distribution to simulate a random process—say to test an inventory control plan, where demands for items arrive randomly but at a known average rate—then what we want is a program that generates an appropriate waiting time for the next event to occur. Suppose the program generates waiting times τ in such a manner that the {probability that τ will be less than T} is $P(T)$. But τ can be less that T only if at least one event occurs in time T, and the probability of this happening is $1 - e^{-nt}$. Thus, $P(T) = 1 - e^{-nt}$, and, from Section 4.4, this kind of distribution can be achieved by the relation $1 - e^{-n\tau} = x$, where x is a random number between 0 and 1. Thus, $\tau = -(1/n) \ln(1 - x)$, or, since $1 - x$ and x have the same distribution,

$$\tau = -\frac{1}{n} \ln x . \tag{4.6}$$

For $n = 4$ the random number generator of Section 4.1 was used in Equation (4.6) to produce a number of time intervals between events. A typical sequence of 100 intervals is given in Table 4.3. The sum of these time intervals is 26.98 seconds. The shortest time between events was .00346 seconds; the longest was 1.723 seconds.

Again choose $n = 4$ (the average number of events per second). An interesting test of the τ-generator described in the last paragraph is to count how many events occur in each successive 2.5-second interval and to list the number of intervals

Table 4.3.   Time intervals between events in a Poisson trial with $n = 4$*

.15504	.20518	.10083	.35879	.05356	.93925	.03963
.73085	.06604	.32313	.65139	.05002	.25720	.17017
.06632	.26587	.16665	.86907	.24839	.20181	.43811
.45651	.28394	.01716	.32305	.39118	.71504	.11934
.07029	.17951	1.03381	.28573	.14876	.29340	.11357
.04311	.12685	.00346	.26142	1.72328	.04240	.04621
.12639	.12371	.38445	.46342	.04733	.18348	.10688
.32269	.19302	.21415	.59306	.17381	.01434	.34074
.18567	.49013	.08507	.06006	.07618	.16682	.40043
.04699	.05108	.01857	.04013	.40855	.10886	.08947
.57691	.08978	.44273	.49538	.24028	.47316	.28309
.14627	.49500	.25211	.79989	.32807	.05458	.46914
.33371	.13325	.05044	.27856	.01894	.04887	.12118
.08830	.00768	.31077	.11271	.60642	.27802	.30306
.28604	.30214					

*Read row by row.

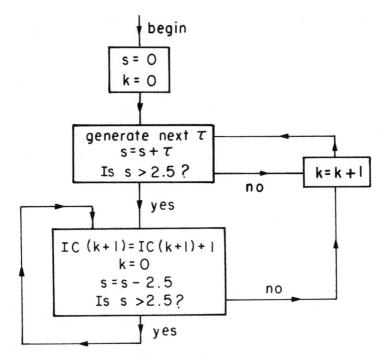

Figure 4.3. Flow chart for Poisson test.

with 0, 1, 2, 3, ... counts. A portion of a tricky little flow chart for a program that does this is shown in Figure 4.3, and a typical result obtained from the corresponding program is given in Table 4.4.

4.6 NORMAL DISTRIBUTION

With each random number x chosen from a uniform distribution in $(0, 1)$, associate a number $z = x - \frac{1}{2}$, so that the numbers z will be uniformly distributed in $(-\frac{1}{2}, \frac{1}{2})$. Now add n of the z-numbers together and divide by n to obtain their average, denoted by y. What will the distribution of the y-numbers look like? For $n = 12$, as a simple test, 500,000 of the y-numbers were generated and the numbers lying in each .01 interval were counted. Figure 4.4 shows the result in the region near $y = 0$. As expected, the peak of the distribution is near $y = 0$, and it falls off fairly rapidly as y moves away from 0.

This kind of bell-shaped curve is characteristic of a *normal distribution*. More precisely, we say that a random variable y has a normal distribution with mean m if the probability $p(y)\,dy$ that y lies in the interval $(y, y + dy)$ is given by

Table 4.4   Number of 2.5-second intervals experiencing various numbers of events, out of a total of 10,000 successive intervals

$j = 0$	0 (0)	$j = 9$	1269 (1251)	$j = 18$	71 (71)
$j = 1$	2 (5)	$j = 10$	1244 (1251)	$j = 19$	35 (37)
$j = 2$	22 (23)	$j = 11$	1127 (1137)	$j = 20$	16 (19)
$j = 3$	76 (76)	$j = 12$	895 (948)	$j = 21$	14 (9)
$j = 4$	199 (189)	$j = 13$	735 (729)	$j = 22$	8 (4)
$j = 5$	355 (378)	$j = 14$	516 (521)	$j = 23$	2 (2)
$j = 6$	609 (631)	$j = 15$	358 (347)	$j = 24$	0 (1)
$j = 7$	944 (901)	$j = 16$	224 (217)	$j = 25$	0 (0)
$j = 8$	1152 (1126)	$j = 17$	126 (128)	$j = 26$	1 (0)

Predicted number from Table 4.2 is shown in parentheses.

$$p(y) \, dy = \frac{1}{\sqrt{2\pi}\sigma} \exp\left(-\frac{(y - m)^2}{2\sigma^2}\right) dy \,, \tag{4.7}$$

where σ is a positive constant. Here y may take any value in the range $(-\infty, \infty)$. Since $p(y)$ is symmetrical about the point $y = m$, it is clear from Equation (4.7) that the average $\bar{y}$ of values of y generated in accordance with Equation (4.7) will be m. Since each time the number y occurs it must have *some* value, it follows that if we add up all possible probabilities, we must get 1; this requires

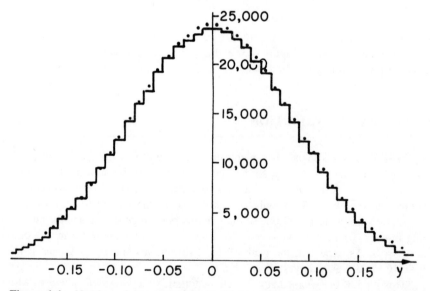

Figure 4.4.   Number of averages lying in .01 interval:—Number from approximating normal distribution: •

$$\int_{-\infty}^{\infty} p(y) \, dy = 1 \tag{4.8}$$

Problem 4.13 asks that Equation (4.8) be verified for the case in which $p(y)$ is given by Equation (4.7). With probability distribution density $p(y)$, the squared deviation from the mean will be $(y - m)^2$; if we add up all of those contributions, we get a formula for the variance, and this turns out to be just σ^2 (hence the notation). Problem 4.13 also asks the reader to show that

$$\int_{-\infty}^{\infty} p(y) y^2 \, dy = \sigma^2 \, . \tag{4.9}$$

We will shortly observe that the results of Figure (4.4) are in accordance with a particular normal distribution. First, however, we quote without proof* a remarkable result known as the *Central Limit Theorem* of probability theory: Let z_1, z_2, $\ldots$, z_n be random variables with probability density functions $p_1(z)$, $p_2(z)$, $\ldots$, $p_n(z)$ (i.e., the probability that z_1 will take a value between z and $z + dz$ is $p_1(z) \, dz$, etc.), each having mean zero. Let σ^2 be the variance of the sum $s = z_1 + z_2 + \cdots + z_n$.

Then as $n \to \infty$, the probability that $s/\sigma < t$ (for any chosen value of t) approaches

$$\frac{1}{\sqrt{2\pi}} \int_{-\infty}^{t} e^{-\xi^2/2} \, d\xi \, . \tag{4.10}$$

This result constitutes the Central Limit Theorem.

Now let us apply this result to predict the outcome of the averaging experiment described at the beginning of this section. Here we are adding 12 z-quantities, each with mean zero and uniformly distributed over the interval $(-\frac{1}{2}, \frac{1}{2})$. The variance of any one of them is the expected value of z^2, which is obtained by averaging z^2 over the interval:

$$E(z^2) = \int_{-1/2}^{1/2} z^2 \, dz = \tfrac{1}{12} \, .$$

We next need the variance of their sum s:

$$\sigma^2 = E(s^2) = E[(z_1 + z_2 + \cdots + z_n)^2]$$

$$= E(z_1^2 + z_2^2 + \cdots + z_n^2 + 2z_1z_2 + \cdots + 2z_{n-1}z_n)$$

$$= E(z_1^2 + z_2^2 + \cdots + z_n^2) \, ,$$

---

*A direct proof can be constructed by the use of Fourier transforms. See, for example, G. Carrier, M. Krook, and C. Pearson, 1966, *Functions of a Complex Variable*, McGraw-Hill, New York, p. 330. The Central Limit Theorem has a long history, dating back almost two centuries.

since, because of the independence of the z_i, $E(z_i z_j) = 0$ for $i \neq j$. Thus,

$$\sigma^2 = \left(\frac{1}{12} + \cdots + \frac{1}{12} \right) = \frac{n}{12} ,$$

and from formula (4.10) we conclude that, if n is reasonably large, the probability that $s/\sqrt{n/12} < t$ is approximately equal to

$$\frac{1}{\sqrt{2\pi}} \int_{-\infty}^{t} e^{-\xi^2/2} \, d\xi .$$

This is equivalent to the statement that the probability that $s/\sqrt{n/12}$ lies between t and $t + dt$ is given by

$$\frac{1}{\sqrt{2\pi}} e^{-t^2/2} \, dt .$$

But this then is the probability that s itself is between $\sqrt{n/12} \, t$ and $\sqrt{n/12} \, (t + dt)$, or that the average, s/n, is between $t/\sqrt{12n}$ and $(t + dt)/\sqrt{12n}$. In the first paragraph we denoted the average by y; the probability $p(y) \, dy$ that y is between y and $y + dy$ is obtained by setting $y = t/\sqrt{12n}$, and we find that

$$p(y) \, dy = \frac{1}{\sqrt{2\pi}} e^{-6ny^2}(\sqrt{12n} \, dy)$$

$$= \sqrt{\frac{6n}{\pi}} e^{-6ny^2} \, dy . \qquad (4.11)$$

In our case we took $n = 12$, and our interval dy was .01. The circles in Figure 4.4 are obtained from Equation (4.11) with $n = 12$ and $dy = .01$; it is seen that the agreement is excellent.

In fact, the averaging process that led to Figure 4.4 is often used as a computer method for generating random numbers with a normal distribution. If numbers w distributed normally with mean m and variance σ^2 are desired, then the relation between w and y is simply

$$w = (\sqrt{12n} \, \sigma)y + m . \qquad (4.12)$$

In practice, the choice $n = 12$ is common. This method, although fast and accurate enough for many purposes, is not exact. An exact method could be constructed by means of Section 4.4, but in the present case the results are somewhat cumbersome; an easier method is described in Problem 4.17.

Quantities obtained experimentally (such as the results of measurements or sampling of population statistics) often involve many uncertainties or errors. Since the errors usually posses some degree of randomness (as likely positive as negative, for example), repeated experiments can be expected to yield results possessing an approximate normal distribution about some mean value. As one example, consider the heights of 20-year-old men; these heights are the end results of random genetic and environmental factors, and one would not be surprised to find them normally distributed about a mean value. In any event, when little is known about the factors involved in the end value of some quantity, a postulated normal distribution is reasonable.

ANNOTATED BIBLIOGRAPHY

M. Abramowitz and I. A. Stegun, eds., 1964, *Handbook of Mathematical Functions*, National Bureau of Standards, Applied Mathematics Series 55, U.S. Government Printing Office, Washington, D.C., 1046p. (Dover reprint available.)

Section 26 summarizes statistical functions and gives many tables. Section 26.8 deals with numerical methods and gives a variety of ways for generating random numbers possessing particular distributions.

B. Carnahan, H. A. Luther, and J. O. Wilkes, 1969, *Applied Numerical Methods*, John Wiley, New York, 604p.

Chapter 8, dealing with statistical methods, includes a number of programs and examples.

D. E. Knuth, 1969, *Seminumerical Algorithms*, *The Art of Computer Programming*, vol. 2, Addison-Wesley, Reading, Mass., 624p.

This second volume of a justly acclaimed series deals extensively with random sequences, their computer generation, and tests for quality; see Chapter 3. The description of the chi-square test (pp. 35–40) is particularly recommended; the historical comment (p. 49) concerning the honesty of a roulette wheel is amusing.

PROBLEMS

4.1 Use a random number generator (installation − supplied or of the kind outlined in Sec. 4.1) to generate n random numbers in (0, 1). Count how many fall in each subinterval of length .01 for $n = 10^2$, 10^3, 10^4, 10^5. Compute the variance for each case and compare with Equation (4.4) (Using $p = .01$; average over the 100 subintervals to obtain σ^2).

4.2 Read a description of the chi-square test (Knuth, 1969 recommended) and use it in connection with the data obtained in Problem 4.1 to test the randomness of the generator.

4.3 Use the Monte Carlo method (see Sec. 4.2) to determine the shaded area of Figure 4.1b to an estimated accuracy of 5%.

4.4 Rather than use a Monte Carlo method in Problem 4.3, we could have generated a regular uniform mesh of points over the unit square and applied an appropriate test to each such point to see whether or not it fell in the shaded area. Try this and compare the accuracy with the results of Problem 4.3. It is important to remark that in a larger number of dimensions (which in some phase space problems may be much greater than three), a uniform distribution may require an unacceptably large number of points.

4.5 Use a Monte Carlo method to compute the polar moment of inertia (see Sec. 4.2) of the shaded area of Figure 4.1*b* about the origin to an estimated accuracy of 5%.

4.6 Starting with random numbers uniformly distributed in (0, 1), write a program to simulate the tossing of a coin (e.g., x in $(0, \frac{1}{2})$ → heads, x in $(\frac{1}{2}, 1)$ → tails). Determine experimentally the probability of a run of at least r successive heads in n tosses, where $r = 5, 7, 10$, and $n = 100, 1000$. (As a partial check, theory shows that 177 tosses are required to give an even chance of a run of at least seven heads.)

4.7 Starting again with random numbers in (0, 1), write an efficient program to simulate the tossing of two dice (each one a cube with faces marked 1 to 6; this remark is for the benefit of the innocent). In craps, betting on the pass line, one wins if the initial toss is 7 or 11 (sum of the two values) and loses if the sum is 2, 3, or 12. If any other number turns up, one continues rolling and wins if that number turns up before a 7 is thrown; otherwise one loses. Determine experimentally the player's chances of winning.

4.8 Write as efficient a program as you can to shuffle a deck of cards.

4.9 The "random walk" motion of a particle along a line may be simulated as follows. Suppose the particle can occupy any of $n + 1$ positions, denoted by $0, 1, 2, \cdots n$. If the particle is at position j, then its next move will be to position $j + 1$, with probability p, or to position $j - 1$, with probability $1 - p$. (Here p is a constant satisfying the condition $0 < p < 1$.) If the particle strikes either of the two end positions $j = 0$ or $j = n$, it is absorbed there, and the random walk terminates. Write a program to model this kind of motion, and use it to determine the probability that the particle will end up at $j = 0$ if it starts at $j = 27$, where $n = 100$ and $p = .52$.

4.10 Devise a program to simulate the following neutron diffusion model. A neutron starts at the origin and moves a unit distance in a random direction. From its new position it again moves a unit distance in a random direction, and so on. Determine its average squared distance from the origin after 100 moves, repeating the overall experiment as many times as you think necessary.

4.11 Write a Poisson process simulator and carry out an experiment of the kind reported in Tables 4.3 and 4.4.

4.12 To generate random numbers y such that the probability that $y < \alpha$ is $P(\alpha) = \alpha^2$ for $0 < \alpha < 1$, with $P(\alpha) = 1$ for $\alpha > 1$, you can use the general method of Section 4.3. An alternative method for this problem is suggested at the end of that section; prove that it is valid.

4.13 The integrand $e^{-\xi^2}$ occurs in the normal distribution expression of Section 4.6. Show that

(a)
$$\int_0^\infty e^{-\xi^2} \, d\xi = \tfrac{1}{2} \sqrt{\pi} \, , \quad \text{and}$$

(b)
$$\int_0^\infty \xi^2 e^{-\xi^2} \, d\xi = \tfrac{1}{4} \sqrt{\pi} \ .$$

Hint: In (a) start with
$$I = \int_0^\infty e^{-x^2} \, dx = \int_0^\infty e^{-y^2} \, dy,$$

so that
$$I^2 = \int_0^\infty \int_0^\infty e^{-(x^2 + y^2)} \, dx \, dy,$$

and change to polar coordinates. In (b) find
$$\frac{\partial}{\partial \alpha} \int_0^\infty e^{-\alpha \xi^2} \, d\xi,$$

and then set $\alpha = 1$.)

4.14 Use a Monte Carlo method to find the volume of the smaller part of the ellipsoid $x^2 + (y^2/2^2) + (z^2/3^2) = 1$ cut off by the plane $x + y + z = \tfrac{1}{2}$.

4.15 Suppose that it is desired to generate random numbers in the finite interval (a, b) with probability density function $p(y)$. Then an *acceptance–rejection* method is to proceed as follows. First, choose a number y lying randomly in (a, b) and compute $\alpha = p(y)/p_m$, where p_m is the maximum of $p(y)$ in (a, b). Next, choose a random number x in $(0, 1)$; if $x < \alpha$, accept y; otherwise start again with a new y-number and a new x. Show that the method works and that its efficiency (the average proportion of numbers y that get accepted) is $1/(b - a) p_m$.

4.16 Write a subroutine to generate the random number of events occurring in each of a sequence of time intervals (each of the same length) if the process is governed by a Poisson distribution witn n events occurring, on the average, per unit of time.

4.17 If y_1 and y_2 are a pair of random numbers, each drawn from a normal distribution with mean 0 and variance 1, then the joint probability that y_1 lies in the range $(y_1, y_1 + dy_1)$ and y_2 lies in the range $(y_2, y_2 + dy_2)$ is

$$\left(\frac{1}{\sqrt{2\pi}} e^{-y_1^2/2} \, dy_1 \right) \left(\frac{1}{\sqrt{2\pi}} e^{-y_2^2/2} \, dy_2 \right) = \frac{1}{2\pi} e^{-(1/2)(y_1^2 + y_2^2)} \, dy_1 \, dy_2 \ .$$

In polar coordinates (r, θ) in the (y_1, y_2) plane, with the area element $r \, dr \, d\theta$, this quantity becomes $(1/2\pi) e^{-r^2/2} r \, dr \, d\theta$. This suggests the following method for generating pairs of normally distributed variables y_1 and y_2: (1) Choose x_1 and x_2 randomly in $(0, 1)$. (2) Define $r = \sqrt{-2 \ln x_1}$ (so that $x_1 = e^{-r^2/2}$, which means that the probability r will lie in $(r, r + dr)$ is $e^{-r^2/2} r \, dr$). (3) Then form $y_1 = r \cos 2\pi x_2$, $y_2 = r \sin 2\pi x_2$. Complete the above argument and use the method to write a subroutine that generates pairs of "normal deviates" corresponding to a mean m and standard deviation σ.

4.18 Two points are chosen randomly along a rod, and the rod is cut at those two points. Determine experimentally the probability that the three pieces of rod so obtained can be used for the sides of a triangle. (This problem is not hard to solve analytically.)

4.19 Write a program to investigate the sensitivity of the solution of a linear equation set (cf. Chapter 2) to small random perturbations of the coefficients.

4.20 A string may be considered to be made up of a number of individual pieces, each of the same length, whose individual breaking strengths are distributed normally about some mean value. Devise a program to estimate the breaking strength of the overall string in appropriate statistical form. Devise some examples.

4.21 During a portion of the day, buses are found to pass a cetain point quite randomly, but with a certain average frequency. A traffic engineer suggests that the expected waiting time for a prospective passenger, arriving randomly at that point, should be one-half of the average time between buses. Assuming a Poisson distribution for the buses, devise a computer experiment to evaluate this suggestion.

INTERPOLATION AND APPROXIMATION

Historically, one of the main incentives for the development of polynomial approximation arose from the desire to interpolate in a table of functional values. Since functional values are now usually obtained, as needed, by direct computer generation, the cumbersome interpolation formalism that at one time embellished textbooks is now of less interest. However, the use of a polynomial to provide a local approximation to a function has become an essential tool in numerical analysis, and many numerical integration, curve-fitting, and differential equation algorithms are based on it.

This chapter discusses polynomial approximation in general and makes applications to curve fitting (e.g., splines). Some of the classical orthogonal polynomials are encountered. Subsequent chapters utilize polynomial approximation in other contexts.

5.1 POLYNOMIAL INTERPOLATION

Suppose that values of a function $f(x)$ are known only for certain discrete values of x, say, for example x_1, x_2, and x_3, and it is desired to estimate the value of $f(x)$ at some other point x_0. This kind of situation, illustrated in Figure 5.1, might arise in connection with tabular data (e.g., the enthalpy of a gas might be tabulated for $T = \ldots$ 1000°C, 1010°C, 1020°C, $\ldots$, and the value at 1014.7°C might be desired) or in the numerical solution of a differential equation that yields values at certain mesh points only. In Figure 5.1, x_0 lies between x_1 and x_2, so the simplest thing to do would be to connect (x_1, y_1) to (x_2, y_2) by a straight line (we denote $f(x_j)$ by y_j for $j = 1, 2, 3$) and find the value of y on this line that corresponds

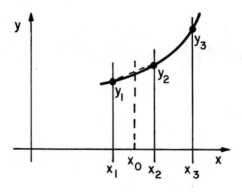

Figure 5.1. Polynomial interpolation.

to x_0. However, this process makes no use of the information provided by the point (x_3, y_3), so we should be able to do better.

A reasonable approach is to "fit" the three points (x_1, y_1), (x_2, y_2), and (x_3, y_3) by a smooth curve, based on the supposition that the unknown function $f(x)$ is itself well behaved, and to find the value of y along this curve for which $x = x_0$. In many situations an nth-order polynomial $y = p_n(x)$ is a suitable choice for such a curve. Polynomials are certainly smooth, are easy to evaluate (and to differentiate), and, moreover, the coefficients of $p_n(x)$ can be determined readily so as to make $p_n(x)$ pass through the given points. In fact, a classical formula due to Lagrange provides the solution almost at once. In the present case we write

$$p_2(x) = y_1 \frac{(x - x_2)(x - x_3)}{(x_1 - x_2)(x_1 - x_3)} + y_2 \frac{(x - x_1)(x - x_3)}{(x_2 - x_1)(x_2 - x_3)}$$

$$+ y_3 \frac{(x - x_1)(x - x_2)}{(x_3 - x_1)(x_3 - x_2)}. \tag{5.1}$$

Observe first that the denominator in each term is some constant (we assume x_1, x_2, x_3 are all different), and that the numerator of each term when multiplied out is a polynomial (here a quadratic), so that, overall, $p_2(x)$ is indeed a polynomial. Moreover, it passes through each of the three points. For at $x = x_1$, for example, the second and third terms vanish, and the first reduces to y_1; a similar calculation can be made at $x = x_2$ and at $x = x_3$. It is noteworthy that this is the *only* second-order polynomial that passes through the three points, for if there were another one, say $q_2(x)$, then $p_2(x) - q_2(x)$ would be a second-order polynomial vanishing at each of x_1, x_2, x_3. However, a second-order polynomial can have three zeros only if it is identically zero (it is shown in Appendix C that an nth-order polynomial has exactly n zeros), so we deduce that $p_2(x) \equiv q_2(x)$.

The general case works just as easily. For n given points, (x_1, y_1), $(x_2,$

y_2), ... (x_n, y_n), an $(n - 1)$st order polynomial passing through these points is given by *Lagrange's interpolation formula*

$$p_{n-1}(x) = y_1 \frac{(x - x_2)(x - x_3) \cdots (x - x_n)}{(x_1 - x_2)(x_1 - x_3) \cdots (x_1 - x_n)}$$

$$+ y_2 \frac{(x - x_1)(x - x_3)(x - x_4) \cdots (x - x_n)}{(x_2 - x_1)(x_2 - x_3)(x_2 - x_4) \cdots (x_2 - x_n)}$$

$$+ \cdots + y_n \frac{(x - x_1)(x - x_2) \cdots (x - x_{n-1})}{(x_n - x_1)(x_n - x_2) \cdots (x_n - x_{n-1})}. \tag{5.2}$$

In Problem 5.1 the reader is asked to write a computer subroutine that evaluates $p_{n-1}(x_0)$ for a given point x_0, where p_{n-1} is obtained from Equation (5.2). [Note that it may be efficient to compute the product $Q = (x_0 - x_1)(x_0 - x_2) \cdots (x_0 - x_n)$ and then to obtain the numerator in the ith term of Equation (5.2) from $Q/(x_0 - x_i)$] The reader is also asked to use this program to interpolate the well-behaved function $f(x) = (1 + 25x^2)^{-1}$ over the range $(-5, 5)$, using n evenly spaced interpolation points. It will turn out that for large n (e.g., $n > 10$) the interpolating polynomial tends to have large oscillations between interpolation points, especially near the ends. This example was given by Runge long before the days of digital computers. It illustrates the fallacy of the remark, "Well, if four interpolation points work so well, let's use 20 points and get a much better result." The basic reason for the trouble arising in the Runge example is that all one is guaranteed by Equation (5.2) is that $p_{n-1}(x)$ passes through the interpolation points; nothing is said about the convolutions that $p_{n-1}(x)$ has to go through in order to do so. The moral is that one should use relatively few interpolation points, since low-order polynomials (quadratics, cubics, etc.) are rather disciplined functions. If a given function $f(x)$ is to be interpolated over a large x-interval L, it may be advantageous to break L up into a number of subintervals and to use different (low-order) interpolation polynomials in different subintervals.

We remark that a formal expression for the discrepancy between $f(x)$ and $p_{n-1}(x)$, for any chosen value of x, can be obtained (see Prob. 5.2). We remark also that $p_{n-1}(x)$ could turn out to have order less than $n - 1$ in some cases. For example, if $f(x)$ had exactly the same value at each of $x_1, x_2, \ldots, x_n$, then $p_{n-1}(x)$ would simply reduce to a constant. A final remark is that in Equation (5.2) the x_i can be given in any order along the x-axis; there is no requirement that $x_1 < x_2 < x_3 \ldots$. We could, for example choose, $x_1 = 2$, $x_2 = 1$, $x_3 = 3$, $x_4 = 0$, and so forth.

Lagrange's interpolation formula can be generalized for use in problems involving two or more dimensions; for the idea see Problem 5.16. A generalization to agreement in slope, as well as value, at each interpolation point is also possible (see Prob. 5.21).

5.2 NEWTON INTERPOLATION

There is a useful alternative to Equation (5.2). Again suppose that we are given a set of n interpolation points $(x_1, y_1), \ldots, (x_n, y_n)$, with the x_i all distinct, through which we want to pass a polynomial $p_{n-1}(x)$. If there were only one point, the solution would be

$$p_0(x) = y_1 .$$

If there were two points, we could write

$$p_1(x) = y_1 + (x - x_1) \alpha_1 ,$$

where α_1 is an appropriate constant. The factor $x - x_1$ ensures that, no matter what α_1 is, $p_1(x)$ passes through (x_1, y_1); we then choose α_1 so that $p_1(x)$ also passes through (x_2, y_2). [In fact, $\alpha_1 = (y_2 - y_1)/(x_2 - x_1)$.] With α_1 chosen in this way, we can now write

$$p_2(x) = y_1 + (x - x_1) \alpha_1 + (x - x_1)(x - x_2) \alpha_2,$$

which, for any α_2, passes through both (x_1, y_1) and (x_2, y_2) and with the appropriate choice for α_2 will also pass through (x_3, y_3). Determining the α_i in sequence in this manner, we obtain finally the *Newton interpolation polynomial*

$$p_{n-1}(x) = y_1 + (x - x_1) \alpha_1 + (x - x_1)(x - x_2) \alpha_2$$
$$+ \cdots + (x - x_1)(x - x_2) \cdots (x - x_{n-1}) \alpha_n .$$

It has become customary to denote α_1 by the expression $[x_1, x_2]$ (which recognizes the fact that data at each of x_1, x_2 must be used to determine α_1), α_2 by $[x_1, x_2, x_3]$, and so on, so that a more common form for $p_{n-1}(x)$ is

$$p_{n-1}(x) = y_1 + (x - x_1)[x_1, x_2] + (x - x_1)(x - x_2)[x_1, x_2, x_3]$$
$$+ \cdots + (x - x_1)(x - x_2) \cdots (x - x_{n-1})[x_1, x_2, \ldots, x_n] .$$

$$(5.3)$$

Had we listed the points x_i in a different order, say $x_3, x_1, x_2, \ldots$, for example, then the first part of Equation (5.3) would have changed from

$$y_1 + (x - x_1)[x_1, x_2] + (x - x_1)(x - x_2)[x_1, x_2, x_3]$$

to

$$y_3 + (x - x_3)[x_3, x_1] + (x - x_3)(x - x_1)[x_3, x_1, x_2] .$$

Each of these is a second-order polynomial through the three points (x_1, y_1), $(x_2,$

y_2), (x_3, y_3); since such a polynomial is unique (cf. Sec. 5.1), these two polynomials are identical. But then the coefficients of the highest powers of x must be the same for each expression, so that $[x_1, x_2, x_3] = [x_3, x_1, x_2]$. An analogous derivation may be made in general, so that the order of the x_i *in any coefficient is immaterial.*

A further remarkable revelation awaits. Suppose we pass a polynomial through the three points (x_1, x_2, x_3), using first the order x_1, x_2, x_3 and then the order (x_2, x_3, x_1). We get the same polynomial $p_2(x)$ in each case, so that

$$y_1 + (x - x_1)[x_1, x_2] + (x - x_1)(x - x_2)[x_1, x_2, x_3]$$
$$= y_2 + (x - x_2)[x_2, x_3] + (x - x_2)(x - x_3)[x_2, x_3, x_1]$$

or, using $[x_2, x_3, x_1] = [x_1, x_2, x_3]$,

$$(x - x_2)\{(x - x_1) - (x - x_3)\}[x_1, x_2, x_3]$$
$$= (y_2 - y_1) + (x - x_2)[x_2, x_3] - (x - x_1)[x_1, x_2] . \qquad (5.4)$$

The left-hand side may be rewritten

$$(x - x_2)(x_3 - x_1)[x_1, x_2, x_3] .$$

But each side of Equation (5.4) is a polynomial (here of first order) in x, and the only way in which two polynomials can be equal for all values of x is if they are identical (as was shown in Sec. 5.1). In particular, the coefficient of x must be the same on the two sides, so that

$$(x_3 - x_1)[x_1, x_2, x_3] = [x_2, x_3] - [x_1, x_2]$$

or

$$[x_1, x_2, x_3] = \frac{[x_2, x_3] - [x_1, x_2]}{x_3 - x_1} . \qquad (5.5)$$

A similar result holds for all of the coefficients in the Newton form of an interpolating polynomial. The reader should prove, for example, that

$$[x_1, x_2, x_3, x_4] = \frac{[x_2, x_3, x_4] - [x_1, x_2, x_3]}{x_4 - x_1} . \qquad (5.6)$$

These results explain the use of the term *divided differences* for these coefficients.

To illustrate the use of the result typified by Equations (5.5) and (5.6), consider the problem of passing a polynomial through the four points (.5, 2), (1, 1), (1.25, .8), (2.5, .4) (these points happen to lie on the curve $y = 1/x$). We form a divided difference table, working in columns from left to right, as shown in Table 5.1. The reader should note carefully the way in which the numerators and denominators are formed.

Table 5.1. Divided differences

x_i	y_i	$[x, x]$	$[x, x, x]$	$[x, x, x, x]$
.5	2			
		$\dfrac{1 - 2}{1 - .5} = -2$		
1	1		$\dfrac{-.8 - (-2)}{1.25 - .5} = 1.6$	
		$\dfrac{.8 - 1}{1.25 - 1} = -.8$		$\dfrac{.32 - 1.6}{2.5 - .5} = -.64$
1.25	.8		$\dfrac{-.32 - (-.8)}{2.5 - 1} = .32$	
		$\dfrac{.4 - .8}{2.5 - 1.25} = -.32$		
2.5	.4			

The desired polynomial is obtained by reading off the coefficients from the top diagonal of Table 5.1:

$$p_3(x) = 2 + (x - .5)(-2) + (x - .5)(x - 1)(1.6)$$
$$+ (x - .5)(x - 1)(x - 1.25)(-.64) .$$

To evaluate $p_3(x)$ for any prescribed value of x, a more efficient numerical procedure would be to rewrite this equation as

$$p_3(x) = 2 + (x - .5)\{-2 + (x - 1)[1.6 + (x - 1.25)(-.64)]\} , \quad \textbf{(5.7)}$$

since this revised form requires fewer multiplications.

If a new point is to be added, and if the Lagrange form (5.2) is being used, every term has to be calculated afresh. The Newton form has the advantage that all previous results are still valid. For example, to add the interpolation point (2, .5) to the above example, we need only adjoin one lower diagonal line of quantities.

The Newton form is easily extended to the case in which slopes are specified at one or more of the data points. To illustrate the idea, suppose that we require $p_4(x)$ to pass through the four points of Table 5.1 and to have slope $-.16$ at the point (2.5, .4). Then we repeat the entry (2.5, .4) in the first two columns; now instead of using $(.4 - .4)/(2.5 - 2.5) = 0/0$ in the $[x, x]$ column, we write the limiting value of a first divided difference as the two points coalesce—that is, the derivative at the point, or $-.16$ in this case. The remainder of the added lower diagonal is completed in the usual way.

The subroutine NPOLY uses the Newton interpolating polynomial to find the value at a given point of the polynomial determined by a set of N points. It is assumed that the x-coordinates of these points are distinct. The reader should observe that, by working with the lower-diagonal coefficients instead of the upper-diagonal ones in the generalization of Table 5.1, array index calculations are minimized.

As pointed out in Problem 5.16, polynomial interpolation in two or more dimensions may be achieved by repeated application of NPOLY.

```
C   SUBROUTINE NPOLY(N,X,Y,C,D,XX,YY) PASSES A POLYNOMIAL THROUGH
C   THE N POINTS (X(I),Y(I)), I=1,2,...,N, AND COMPUTES THE VALUE
C   YY OF THIS POLYNOMIAL AT THE POINT XX.  THE NEWTON FORM OF THE
C   INTERPOLATING POLYNOMIAL IS USED.  THE GIVEN X(I) VALUES MUST
C   BE DISTINCT.
C
C   INPUT:
C     N = NUMBER OF INTERPOLATION POINTS
C     X = ARRAY OF X-VALUES, DIMENSION N (VALUES MUST BE DISTINCT)
C     Y = ARRAY OF Y-VALUES, DIMENSION N
C     C = WORKING ARRAY, DIMENSION N
C     D = WORKING ARRAY, DIMENSION N
C    XX = VALUE OF X AT WHICH VALUE OF POLYNOMIAL IS DESIRED
C
C   OUTPUT:
C    YY = VALUE OF POLYNOMIAL AT POINT XX
C
C   NOTE:   CALLING PROGRAM MUST PROVIDE SPACE FOR ARRAYS C AND D
C
        SUBROUTINE NPOLY(N,X,Y,C,D,XX,YY)
        DIMENSION X(N),Y(N),C(N),D(N)
        DO 6 I=1,N
6       C(I)=Y(I)
        N1=N-1
C   COMPUTE NEWTON COEFFICIENTS
        DO 8 I=1,N1
        J=N-I
        DO 7 K=1,J
7       C(K)=(C(K+1)-C(K))/(X(K+I)-X(K))
8       CONTINUE
C   NOW COMPUTE VALUE OF POLYNOMIAL AT XX
        D(1)=C(1)
        DO 9 I=2,N
9       D(I)=C(I)+(XX-X(I))*D(I-1)
        YY=D(N)
        END
```

5.3 SPLINES

We have seen that polynomial interpolation may be used to obtain approximations to the values (and derivatives) of a function $f(x)$ whose values are known at a set of mesh points x_i. To avoid the oscillations often associated with high-order polynomials, this process is usually restricted to interpolation through a fairly small number of points adjacent to the point of interest. If a function $f(x)$ is to be "fitted" over an interval that includes a large number of data points, then rather than try to find a single polynomial that passes through all of the points, it is usually preferable to either relax this constraint and require the polynomial only to pass "close" to the given points or to use different polynomials to represent different portions of $f(x)$. This section deals with one version of the latter approach, termed *spline interpolation*, in which adjacent polynomials are smoothly connected to one another. The name arises from the draftsman's spline, which is moved from one part of a curve to another, and bent each time for best local fit.

Suppose that a number of points (x_1, y_1), (x_2, y_2), ... , (x_n, y_n) (termed *knots*) are given, with $x_1 < x_2 < x_3, \ldots , < x_n$, and that we want to connect each pair of adjoining points by a polynomial. Although straight lines, parabolas, cubics, quartics, ... , could be used, the most common choice, motivated by a compromise between flexibility and economy of calculation, is to use cubics. Denote the cubic joining (x_i, y_i) to (x_{i+1}, y_{i+1}) by

$$p_i(x) = y_i + c_{i1}(x - x_i) + c_{i2}(x - x_i)^2 + c_{i3}(x - x_i)^3 , \tag{5.8}$$

where we have used the fact that $p_i(x_i) = y_i$ (see Fig. 5.2). Note that we have changed the notation from previous sections in that $p_i(x)$ now denotes the ith polynomial rather than a polynomial of order i. One condition that helps determine the c_{ij} coefficients is that $p_i(x)$ must also pass through the point (x_{i+1}, y_{i+1}). But this imposes only one condition on the three coefficients, so that two more conditions could be satisfied. All together, there are $n - 1$ polynomial sections, so that, beyond the requirement that each $p_i(x)$ pass through its two endpoints, a total of $2(n - 1)$ further conditions can be chosen. A natural choice is to make the slope and curvature continuous at the knots to provide a smooth transition between adjacent cubics. This choice leads to

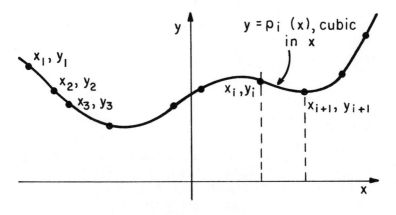

Figure 5.2. Splines.

$$p'_{i-1}(x_i) = p'_i(x_i) , \qquad p''_{i-1}(x_i) = p''_i(x_i) \tag{5.9}$$

for $i = 2, 3, \ldots , n - 1$. There are still two conditions left over. It may of course happen that the slopes at x_1 and x_n are specified as part of the problem data, and this specification would use up the remaining two conditions. If this is not the case, we are free to require the curvature to vanish at each point x_1, x_n (this is called the *free end condition* and leads to *natural* spline interpolation), or to require $p_1(x)$ to be the same cubic as $p_2(x)$ and $p_{n-2}(x)$ to be the same cubic as $p_{n-1}(x)$ (in which case p''' is continuous at x_2 and x_{n-1}; this is termed the *not-a-knot* condition). Other choices are also possible, but the above are the most common.

The cubic $p_i(x)$ between any pair of mesh points x_i and x_{i+1} has four coefficients and so is determined by a specification of its endpoint values and slopes— that is, by y_i, y_{i+1} and s_i, s_{i+1}. In terms of the coefficients of Equation (5.8), we find easily that

$$c_{i1} = s_i ,$$

$$c_{i2} = 3 \frac{y_{i+1} - y_i}{h_i^2} - \frac{2s_i + s_{i+1}}{h_i} ,$$

$$c_{i3} = \frac{s_i + s_{i+1}}{h_i^2} - 2 \frac{y_{i+1} - y_i}{h_i^3} , \tag{5.10}$$

where $h_i = x_{i+1} - x_i$. Computationally, it is often convenient to treat the slopes s_i as the unknowns; once they have been found, Equations (5.10) then determine the individual cubics.

To find the s_i, we write the condition that the curvatures of adjoining cubics must match at their connection point. This condition leads to

$$s_{i-1}\left(\frac{1}{h_{i-1}}\right) + s_i\left(\frac{2}{h_{i-1}} + \frac{2}{h_i}\right) + s_{i+1}\left(\frac{1}{h_i}\right) = \frac{3}{h_i^2}(y_{i+1} - y_i)$$

$$+ \frac{3}{h_{i-1}^2}(y_i - y_{i-1}) \tag{5.11}$$

for $i = 2, 3, \ldots , n - 1$, where $h_i = x_{i+1} - x_i$ as before. To this set of $n - 2$ equations we must adjoin the two end conditions, which may also be phrased in terms of the s_i quantities. From Equations (5.11) it is clear that the overall coefficient matrix is of tridiagonal form, so that the special case of Gaussian elimination outlined in Problem 2.13 is applicable. Thus, the solution process for the s_i is very efficient. An example of spline interpolation (with a partial solution) will be found in Problem 5.7. Note, incidentally, that once the c_{ij} have been determined, an economical way in which to calculate $p_i(x)$ from Equation (5.8) is to write

$$p_i(x) = y_i + (x - x_i)[c_{i1} + (x - x_i)\{c_{i2} + (x - x_i) c_{i3}\}] . \tag{5.12}$$

5.4 MORE ON CURVE (AND SURFACE) FITTING

In this section we collect a number of remarks pertaining to the problems of passing a smooth curve or surface through, or nearly through, a given set of points. This kind of situation arises in computer-aided design (CAD) and also in numerically controlled manufacturing processes. One example is sketched in Figure 5.3, where the solid points represent measurement data for an experimental wing profile transmitted to a shop that has to connect the points with a smooth curve and then generate the resulting profile section.

Complete Curve

It may happen, as in Figure 5.3, that we want to draw a spline through a set of points forming a complete curve with or without natural endpoints. The obvious difficulty is that portions of the curve will become nearly perpendicular to the x-axis, and for these portions polynomials in x are no longer appropriate. One approach is to join each pair of points by a chord, as in Figure 5.4, and to use this chord as a local x'-axis. The local (x', y') system may then be used to obtain a cubic for y' in terms of x', and the conditions of matching slope and curvature may be imposed at junction points as before. The problem is now nonlinear, but it can be solved by the methods of Chapter 3. It is sometimes useful to rescale the original problem (e.g., perhaps the profile in Figure 5.3 could be stretched in the vertical direction prior to application of the spline algorithm).

Alternatives to Splines

Splines are useful but are not always the best choice. When one tries to go around a fairly sharp curve, as in the leading edge portion of the wing profile of Figure 5.3, the cubics constituting the spline may evidence an unacceptable ripple. An alternative approach would be to begin by "filling in" intermediate points according to some kind of smoothness criterion; if this can be done appropriately, and repetitively, then after some passes of the algorithm the set of points will be

Figure 5.3. Wing profile.

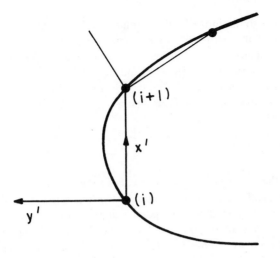

Figure 5.4. Local coordinate system.

dense enough that they can be joined together very simply (even by straight lines!). One effective procedure for filling in points is as follows.

The basic idea is to insert new points so as to minimize ripple in curvature, where the ripple is measured by means of a merit function M. Consider a set of points in the plane, as in Figure 5.5, numbered in sequence from 1 to n. A curvature c_j may be associated with the jth point (not an endpoint) by computing the reciprocal radius of a circle passing through the points $j - 1, j$, and $j + 1$; we then define

$$M = |c_3 - c_2|^\alpha + |c_4 - c_3|^\alpha + \cdots + |c_{n-1} - c_{n-2}|^\alpha ,$$

where α is a positive parameter (typically, $1 \le \alpha \le 2$). Suppose now that a new point is to be inserted between points 3 and 4 in Figure 5.5. Draw the perpendicular bisector to the chord line joining these points, and insert the new point on this

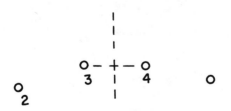

Figure 5.5. Insertion of new points.

bisector line so as to minimize M (the formula for M now includes the new point). Do this for all of the perpendicular bisectors, iterating as necessary until M is minimized. The filled-in points in Figure 5.3 were obtained by repeated use of this algorithm; for this profile the point A at the trailing edge was not included in the formula for M. A second example is given in Figure 5.6, where the solid points were initially specified; it is gratifying that a near-helix resulted.

Surfaces

Consider the rectangular mesh of points in the (x, y) plane shown in Figure 5.7, where a typical mesh point has coordinates (x_i, y_j), $i = 1, 2, \ldots, m$ and $j = 1, 2, \ldots, n$. Suppose that mesh point values of a function $f(x, y)$ are given, and that it is desired to fit $f(x, y)$ with a two-dimensional spline $\phi(x, y)$. For boundary conditions we choose $\partial^2\phi/\partial x^2 = 0$ on the two vertical sides and $\partial^2\phi/\partial y^2 = 0$ on the two horizontal sides; these conditions are analogous to the previous "natural" boundary condition. One approach is to first construct splines in the

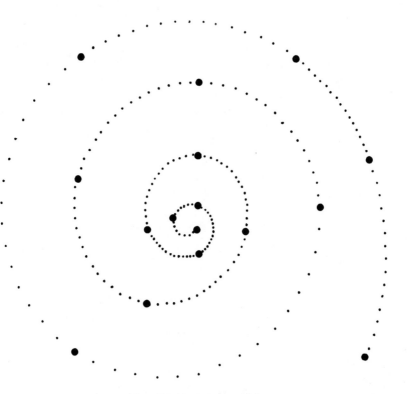

Figure 5.6. Helix example.

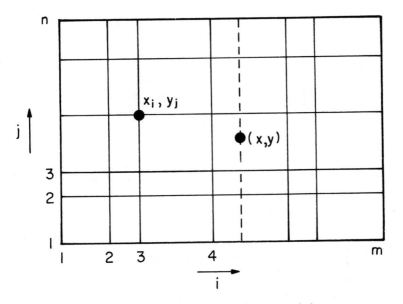

Figure 5.7. Two-dimensional spline interpolation.

usual way along each of the horizontal mesh lines—that is, for fixed values of y equal to $y_1, y_2, \ldots, y_n$, respectively. Then to determine ϕ at some point (x, y), draw a vertical line through (x, y) (dashed line), read off the values of ϕ (as determined by the previous spline calculations) at those points where this line cuts the horizontal spline lines, and use these values to construct a spline curve along the dashed line. Remarkably enough, we would get the same value for $\phi(x, y)$ if the roles of the x and y directions were revised (cf. Prob. 5.9).

Best Fit

In the interpolation discussion to date, whether for polynomials or splines, the approximating function has been required to pass through each data point. However, this is often an unnecessarily restrictive assumption, and, from a practical point of view, the data points are often subject to experimental error anyway. Although the following discussion could be phrased equally well in terms of splines, we choose a single polynomial for illustrative purposes. Suppose that a number of points $(x_1, y_1), (x_2, y_2), \ldots, (x_n, y_n)$ are given in the plane (Fig. 5.8), and we wish to pass a polynomial approximately through these points. We choose here a parabola of the form $p(x) = a + bx + cx^2$. How do we choose the coefficients a, b, c for best fit to the given points? (This problem could arise if we wanted to represent some experimental results by a simple empirical formula.) The problem is, of course, to minimize in some way the average distance of the points from the curve

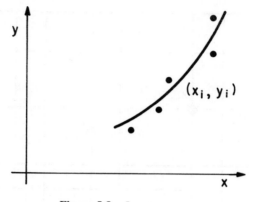

Figure 5.8. Least squares.

$y = p(x)$. The use of squares is analytically convenient, and a common approach is to try to choose the coefficients a, b, c so as to minimize

$$S = \sum_{i=1}^{n} [y_i - (a + bx_i + cx_i^2)]^2 .$$

The values of a, b, c for which S is a minimum must be such that the partial derivative of S with respect to each of these quantities vanishes, and this condition leads to

$$\sum_{i=1}^{n} [y_i - (a + bx_i + cx_i^2)] = 0 ,$$

$$\sum_{i=1}^{n} x_i [y_i - (a + bx_i + cx_i^2)] = 0 ,$$

$$\sum_{i=1}^{n} x_i^2 [y_i - (a + bx_i + cx_i^2)] = 0 ,$$

which is a set of three linear equations for a, b, c. The case in which $p(x)$ is linear in x (so $y = p(x)$ is a straight line) is frequently encountered in statistics (perhaps using distorted—say, logarithmic—scales for x and y); the subject is there called *regression analysis*. Of course, there is no need to restrict oneself to polynomials; exponential or other functions can be useful.

5.5 ORTHOGONAL POLYNOMIALS

There are some special families of polynomials that are very useful in numerical analysis. Usually the members of any one family satisfy a certain "orthogonality" condition, which we will define shortly. There are so many such families that here we will consider only one in detail: the family of Chebyshev polynomials.

We will derive some of their properties and then simply list similar properties for a selection of other families.

To motivate the idea of functional orthogonality, we begin with the familiar three-dimensional vector case. Suppose we have a vector whose components in the x-, y-, z-directions are denoted by f_1, f_2, f_3, respectively, and a second vector whose corresponding components are g_1, g_2, g_3. Then these two vectors are orthogonal to one another if their dot or scalar product vanishes—that is, if $f_1 g_1 + f_2 g_2 + f_3 g_3 = 0$. If we had n dimensions instead of three, with the two vectors having components $f_1, f_2, \ldots, f_n$ and $g_1, g_2, \ldots, g_n$, then the natural generalization of orthogonality is to require $f_1 g_1 + f_2 g_2 + \cdots + f_n g_n = 0$. Now suppose that two functions, $f(x)$ and $g(x)$, are defined over the interval (a, b) and consider $\int_a^b f(x) g(x)$ dz. The definition of an integral tells us that this expression is equal to the limit (as $n = \infty$) of a sum in which we divide the interval $b - a$ into n subintervals Δx_i (with f_i = value of f at some point within Δx_i, etc.):

$$\int fg \, dx = \lim_{n \to \infty} \sum_{i=1}^{n} (f_i g_i) \, \Delta x_i \, .$$

If the Δx_i are all of equal size, then, apart from the common factor Δx, the righthand side has the form of a scalar product in n-dimensional space, and this motivates the definition that two functions $f(x)$ and $g(x)$ are said to be orthogonal over (a, b) if $\int_a^b f(x) g(x) \, dx = 0$. We can generalize a bit and introduce a positive *weight function* $w(x)$, which simply biases the various portions of the interval differently, and say that f and g are orthogonal with respect to the weight function $w(x)$ if $\int_a^b w(x) f(x) g(x) \, dx = 0$.

Now let us define the nth Chebyshev polynomial $T_n(x)$ (the T comes from the spelling Tschebycheff). If n is any integer, then we will show that cos $n\theta$ can be expressed as a polynomial in cos θ; with cos θ replaced by x, this polynomial gives us $T_n(x)$. The first few are obtained from standard trigonometric identies:

$$\cos(0 \cdot \theta) = 1 \Rightarrow T_0(x) = 1 \, ,$$

$$\cos(1 \cdot \theta) = \cos \theta \Rightarrow T_1(x) = x \, ,$$

$$\cos 2\theta = 2 \cos^2 \theta - 1 \Rightarrow T_2(x) = 2x^2 - 1 \, ,$$

$$\cos 3\theta = 4 \cos^3 \theta - 3 \cos \theta \Rightarrow T_3(x) = 4x^3 - 3x \, .$$

More generally, we have

$$\cos(n + 1) \theta + \cos(n - 1) \theta = 2 \cos n\theta \cdot \cos \theta \, ,$$

which shows (by recursion starting with $n = 1$) that cos $n\theta$ can indeed be expressed as a polynomial in cos θ, and which also gives the *recursion formula*

$$T_{n+1}(x) + T_{n-1}(x) = 2xT_n \, . \tag{5.13}$$

Using Equation (5.13), we find the next few $T_n(x)$ to be

$$T_4(x) = 8x^4 - 8x^2 + 1 ,$$
$$T_5(x) = 16x^5 - 20x^3 + 5x ,$$
$$T_6(x) = 32x^6 - 48x^4 + 18x^2 - 1 .$$

It follows easily from Equation (5.13) that the T_n of odd order are odd functions of x [i.e., involve only odd powers of x, so that $T_n(-x) = -T_n(x)$], and that the T_n of even order are even functions of x [so that $T_n(-x) = T_n(x)$]. It also follows from Equation (5.13) that the coefficient of x^n in T_n equals 2^{n-1} for $n \geq 1$.

To make the relation between θ and x in $\cos \theta = x$ more specific, choose the range of θ to be from $-\pi$ to 0; correspondingly, x ranges from -1 to 1. Since $\cos n\theta$ vanishes in the interval $(-\pi, 0)$ at the n points

$$\theta_k = -\frac{(2k - 1)\pi}{2n} , \qquad k = 1, 2, \ldots, n ,$$

it follows that $T_n(x)$ vanishes at the n points

$$x_k = \cos \frac{(2k - 1)\pi}{2n} , \qquad k = 1, 2, \ldots, n . \tag{5.14}$$

Thus all the zeros of T_n are real and distinct and lie in the interior of the interval $(-1, 1)$. The situation is depicted geometrically in Figure 5.9. An interesting observation, which follows from a study of this figure, is that the zeros of $T_{n+1}(x)$ "interlace" those of $T_n(x)$; that is, between each pair of adjacent zeros of T_n lies

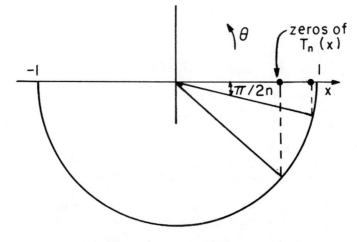

Figure 5.9. Zeros of $T_n(x)$.

exactly one zero of T_{n+1}. Another observation is that, as n increases, the zeros of T_n tend to cluster more tightly near the two endpoints of the interval $(-1, 1)$.

A result that follows from $T_n(x) = \cos n\theta$, with $\cos \theta = x$, is that $T_n(x) = 1$ and $T_n(-1) = (-1)^n$ for all n. Note also that $|T_n(x)| \le 1$ for all x in $(-1, 1)$. This last result is sometimes used in a process called *Chebyshev economization*, the idea of which will become clear from a simple example. Over the interval $(-1, 1)$ we can write

$$\sin x \cong x - \frac{x^3}{3!} + \frac{x^5}{5!} - \frac{x^7}{7!} \tag{5.15}$$

with an error no greater than $1/9! = (2.76)(10^{-6})$. But we can also express x^7 as

$$x^7 = \frac{1}{64} [T_7 + 7T_5 + 21T_3 + 35T_1] ,$$

so that an equivalent form for Equation (5.15) is

$$\sin x \cong x - \frac{x^3}{3!} + \frac{x^5}{5!} - \frac{1}{(64)(7!)} [T_7 + 7T_5 + 21T_3 + 35T_1] . \tag{5.16}$$

Now $|T_7| \le 1$ in $(-1, 1)$, so that if we discard the term in T_7 from Equation (5.16), we get an approximation for $\sin x$ involving only powers up to x^5, and which is in error by no more than $1/9! + 1/(64)(7!) = (5.86)(10^{-6})$. [Had we just dropped the term $x/7!$ from Equation (5.15), the error bound would have been $1/7! = (1.98)(10^{-4})$.] Thus, we have "economized" the calculation of $\sin x$ without unacceptably increasing the error. If the interval of interest is not $(-1, 1)$, but rather (a, b), we simply rescale—that is, map the interval (a, b) onto $(-1, 1)$ before applying this process.

Chebyshev polynomials also play a useful role in polynomial interpolation. Problem 5.2 outlines a derivation of the fact that if a function $f(x)$ is interpolated over an interval (a, b) by means of a polynomial $p_{n-1}(x)$, which agrees with $f(x)$ at the points $x_1, x_2, \ldots, x_n$, then at any other point x in this interval the difference between $f(x)$ and $p_{n-1}(x)$ is given by

$$f(x) - p_{n-1}(x) = (x - x_1)(x - x_2) \cdots (x - x_n) \frac{f^{(n)}(\xi)}{n!} , \tag{5.17}$$

where ξ is some point in (a, b). Here $f^{(n)}(\xi)$ denotes the nth derivative of f evaluated at the point ξ. In general, ξ will depend on the choice of x, so we cannot say much about the last factor in Equation (5.17), except that it is bounded by the maximum absolute value of the nth derivative of f in (a, b). However, there is the possibility of minimizing the righthand side of Equation (5.17)—that is, the discrepancy between f and p_{n-1}—by choosing the interpolation points x_i so that the maximum absolute value of $(x - x_1)(x - x_2) \cdots (x - x_n)$, as x ranges over the interval (a, b), is as small as possible.

In order to attack this problem, let us first scale the interval (a, b) so that it becomes $(-1, 1)$. Then we want to minimize $|(x - x_1) \cdots (x - x_n)|$ over $(-1, 1)$ by an appropriate choice of the x_i points. It was first pointed out by Chebyshev that *this can be done by choosing for the x_i the n zeros of $T_n(x)$.* The proof is straightforward. Note first that $(x - x_1)(x - x_2) \cdots (x - x_n)$ is a polynomial of order n, with leading coefficient unity; moreover, if the x_i were the zeros of T_n, then we would have $(x - x_1)(x - x_2) \cdots (x - x_n) = T_n(x)/2^{n-1}$. Thus, what we want to show is that, among all nth order polynomials $q_n(x)$ with leading coefficient unity, $T_n(x)/2^{n-1}$ has the least upper bound for its absolute value over $(-1, 1)$. Now T_n attains its upper bound (unity) at the two ends of the interval and also between each pair of its n zeros (think of the $\cos n\theta$ interpretation of T_n), so that $|T_n/2^{n-1}|$ attains its least upper bound of $1/2^{n-1}$ a total of $n + 1$ times in the (closed) interval. If there were a better polynomial $Q_n(x)$ of the kind considered, in the sense that its least upper bound were *less* than $1/2^{n-1}$, then the difference polynomial $T_n/2^{n-1} - Q_n(x)$ would (a) be of order $n - 1$ since the two leading powers cancel, and (b) would have the same sign as $T_n/2^{n-1}$ at those points where $|T_n/2^{n-1}|$ attains its maximum. But this means that the difference polynomial would alternate in sign at $n + 1$ points and so would vanish at n intermediate points, which is not possible for a polynomial of order $n - 1$ unless it is identically zero. This completes the proof and explains why interpolation points are often taken to coincide with the zeros of a Chebyshev polynomial.

Consider next, for n and m integral,

$$I = \int_{-\pi}^{0} \cos m\theta \cos n\theta \, d\theta = \frac{1}{2} \int_{-\pi}^{0} [\cos (m + n)\theta + \cos(m - n)\theta] \, d\theta .$$

We find $I = 0$ for $m \neq n$, $I = \pi$ for $m = n = 0$, and $I = \pi/2$ for $m = n \neq 0$. With the change in variable $x = \cos \theta$, this result becomes

$$\int_{-1}^{1} \frac{1}{\sqrt{1 - x^2}} T_m(x) T_n(x) \, dx = \begin{cases} 0 & \text{for } m \neq n , \\ \frac{1}{2} \pi & \text{for } m = n \neq 0 , \\ \pi & \text{for } m = n = 0 . \end{cases} \quad \textbf{(5.18)}$$

Thus, the Chebyshev polynomials are orthogonal to one another with respect to the weight function $w(x) = (1 - x^2)^{-1/2}$.

There are many other orthogonal polynomials. Denoting the weighting function by $w(x)$ and the interval of integration by (a, b), some of the more common ones are

Legendre Polynomials: $\quad P_n(x), \quad w = 1, \quad (a, b) = (-1, 1) ,$

Hermite Polynomials: $\quad H_n(x), \quad w = e^{-x^2}. \quad (a, b) = (-\infty, \infty) ,$

Laguerre Polynomials: $\quad L_n(x), \quad w = e^{-x} \quad (a, b) = (0, \infty) .$

Each of these polynomials satisfies properties analogous to those derived above for the Chebyshev polynomials. The derivations of a general recurrence relation and of the fact that all the zeros of each polynomial are real and lie in (a, b) are sketched in Problems 5.18 and 5.19.

As a final topic in this section, consider again the problem of approximating a given function $f(x)$ in terms of a set of powers of x. Suppose that over an interval (a, b) we write

$$f(x) \cong a_0 + a_1 x + \cdots + a_n x^n . \tag{5.19}$$

One way in which to determine the coefficients would be to use the least squares idea of Section 5.4 and to choose the a_i to minimize

$$I = \int_a^b [f(x) - (a_0 + a_1 x + \cdots a_n x^n)]^2 \, dx . \tag{5.20}$$

Again, a necessary condition for a minimum is that the derivative of I with respect to each of the a_i should vanish, and this gives a set of n linear equations to determine the a_i. Unfortunately, the coefficient matrix turns out to be ill-conditioned (see Prob. 5.13), so that for n greater than 6 or so this approach becomes difficult. It is much better to first map the interval (a, b) onto the interval $(-1, 1)$ and to then approximate the "mapped" function, say $F(x)$, by a sum of Legendre polynomials:

$$F(x) \cong b_0 + b_1 P_1(x) + \cdots + b_n P_n(x) . \tag{5.21}$$

(The result is essentially equivalent to that of Eq. (5.19), since each P_j is a polynomial in x so that the individual powers of x in Eq. (5.21) could be collected.) The appropriate integral to minimize is now

$$J = \int_{-1}^1 [F(x) - (b_0 + b_1 P_1 + \cdots + b_n P_n)]^2 \, dx .$$

Using the orthogonality of the P_i, the result of the differentiation process will be to give a set of equations of the typical form

$$\int_{-1}^1 F(x) P_j(x) \, dx = b_j \int_{-1}^1 P_j^2(x) \, dx ,$$

which determine the b_j at once. No set of linear equations need be solved. It may be remarked that an approximation of $F(x)$ in terms of Chebyshev polynomials could have some advantages, since the error in

$$F(x) \cong c_0 + c_1 T_1(x) + \cdots + c_n T_n(x) \tag{5.22}$$

is often given, approximately, by the first omitted term [here, $c_{n+1} T_{n+1}(x)$], and since T_{n+1} is bounded by unity in $(-1, 1)$, the error can often be assumed to be of order c_{n+1}.

ANNOTATED BIBLIOGRAPHY

M. Abramowitz and I. A. Stegun, eds., 1964, *Handbook of Mathematical Functions*, National Bureau of Standards, Applied Mathematics Series 55, U.S. Government Printing Office, Washington, D.C., 1046p. (Dover reprint available.)

Many formulas and tables are included. See Sections 22 and 25.

P. J. Davis, 1963, *Interpolation and Approximation*, Blaisdell, Waltham, Mass., 303p. (Dover reprint, 1975.)

The flavor is somewhat theoretical, but this is where to go if you can't find the theorem or result you want elsewhere. The bibliography is valuable, although slightly dated.

C. de Boor, 1978, *A Practical Guide to Splines*, Springer-Verlag, New York, 392p.

Although the notation is a bit cumbersome, Chapters 1 through 4 provide excellent discussions of polynomial interpolation and cubic splines. The computer programs have been carefully designed.

L. Fox and I. B. Parker, 1978, *Chebyshev Polynomials in Numerical Analysis*, Oxford University Press, London, 205p.

This very readable text includes a discussion of applications of Chebyshev approximation methods to the approximate solution of differential equations.

F. Schied, 1968, *Theory and Problems of Numerical Analysis*, Schaum Outline Series, McGraw-Hill, New York, 422p.

Chapter 21 uses the attractive "solved problems" approach to discuss least squares approximations; Chebyshev polynomials are included. Chapter 22 extends the discussion to the case in which the greatest error of the approximation is minimized (the "min-max" idea), and Chapter 23 considers the use of rational functions (ratios of polynomials) as approximating functions.

PROBLEMS

5.1 Given n points in the plane, denoted by $(X(J), Y(J))$ for $J = 1, 2, \ldots, n$, write a subroutine that uses Lagrange interpolation [Eq. (5.2)] to find the value of $p_{n-1}(x)$ (say Y0) at a given point X0. [Hint: Note the remarks following Eq. (5.2)]. Compute the required number of arithmetical operations as a function of n. Next, apply this program to the Runge function $f(x) = 1/(1 + 25x^2)$, using n equally spaced interpolation points in $(-5, 5)$ and plot $p_{n-1}(x)$ for $n = 4, 8, 16$. In each case choose enough points x that $p_{n-1}(x)$ can be adequately graphed. If n is fixed and a large number of x-values are to be used, can you think of a more efficient method for calculating the various $p_{n-1}(x)$ values?

5.2 Obtain an expression for the error term in polynomial interpolation by filling in the details of the following procedure. Let $p_{n-1}(x)$ be the interpolating polynomial to $f(x)$ of order $n - 1$, obtained by use of the distinct interpolation points $x_1, x_2, \ldots, x_n$. Assume $f(x)$ is continuously differentiable at least n times. Write

$$f(x) = p_{n-1}(x) + E(x) ,$$

where $E(x)$ is the error term (the discrepancy between f and p_{n-1}) at any point x; since $E(x) = 0$ for $x = x_1, x_2$, and so forth, we can write

$$E(x) = (x - x_1)(x - x_2) \cdots (x - x_n)g(x) .$$

Let t be any chosen point, and let I be an interval containing t as well as all of the x_i. Form the function

$$F(x) = f(x) - p_{n-1}(x) - (x - x_1)(x - x_2) \cdots (x - x_n)g(t) ,$$

which clearly vanishes at the $n + 1$ points $x_1, x_2, \ldots, x_n, t$ in I. Since the derivative of F must vanish at least once between any two zeros of F, it follows that $F'(x)$ must have at least n zeros in I. Similarly, F'' must have at least $n - 1$ zeros in I. Continuing, $F^{(n)}$ must have at least one zero in I, say at the point ξ. Taking the nth derivative of $F(x)$, as defined above, we find

$$0 = F^{(n)}(\xi) = f^{(n)}(\xi) - n!g(t) ,$$

so that

$$E(x) = (x - x_1)(x - x_2) \cdots (x - x_n)\frac{f^{(n)}(\xi)}{n!} ,$$

where ξ (which depends on x, of course) lies in I.

5.3 One way in which to obtain a polynomial $a_0 + a_1x + a_2x^2 + \cdots + a_{n-1}x^{n-1}$ passing through the n points $(x_1, y_1), (x_2, y_2), \ldots, (x_n, y_n)$ is to solve the set of n simultaneous equations

$$a_0 + a_1x_1 + a_2x_1^2 + \cdots + a_{n-1}x_1^{n-1} = y_1 ,$$

$$\vdots$$

$$a_0 + a_1x_n + a_2x_n^2 + \cdots + a_{n-1}x_n^{n-1} = y_n .$$

How would this approach compare, in terms of computer effort and possible ill-conditioning, with the use of Equation (5.2)?

5.4 Modify the subroutine given in Section 5.2 so as to also determine the derivative of the polynomial at the point XX. [*Note:* this can be done efficiently by use of the D(I) values already found.] Test the result for a nontrivial case in which $N = 5$.

5.5 Low-order polynomial interpolation is frequently used to obtain an approximation to the derivative of a function known only at certain discrete points. Let $f(x)$ be given at x_1, x_2, x_3 (not necessarily equally spaced), where $x_1 < x_2 < x_3$. Pass a parabola through the three points $(x_1, f(x_1))$, $(x_2, f(x_2))$, $(x_3, f(x_3))$, and so find an approximation to $f'(x_1), f'(x_2)$. How does $f'(x_2)$ compare to the slope of the chord joining the outer two points for (a) equal spacing of the x_i points, and (b) unequal spacing of the x_i points? If $f(x)$ happens to be a cubic, find the errors in $f(x_1)$ and $f'(x_2)$.

5.6 A table for the specific enthalpy h of air (British units) as a function of temperature T has the following entries:

T	850	860	870	880	890	900
h	204.01	206.46	209.01	211.66	214.41	217.27

Write a computer subprogram to provide h for any value of T in this range; describe how you chose your interpolation procedure.

5.7 Pass a spline through the five points $(1, 1)$, $(2, \frac{1}{2})$, $(3, \frac{1}{3})$, $(4, \frac{1}{4})$, and $(5, \frac{1}{5})$, all of which lie on the curve $y = 1/x$. Use each of the end conditions (1) $s_1 = -1$, $s_5 = -.04$, (2) $y'' = 0$ at x_1 and x_5, and (3) the "not-a-knot" condition at the ends. *Partial answer:* The first polynomial in each case has the form

(1) $1 - x + .720714x^2 - .220714x^3$,

(2) $1 - .583929x + .0839286x^3$,

(3) $1 - .766667x + .3166667x^2 - .05x^3$.

5.8 Draw a circle and choose a number of equally spaced points on its circumference. Use the complete curve method of Section 5.4 to draw a spline curve through these points, and compare the resulting figure with the original circle. Use 4, 8, and 16 points. (Note that the character of the slope-determination equation is the same at each point.)

5.9 Prove the statement concerning interchange of x and y variables made at the end of the part entitled Surfaces in Section 5.4. [*Hint:* Because of linearity, we need prove the result only for the case in which $f(x_i, y_j) = 0$ for all values of i and j except for one, say $i = I$ and $j = J$; at this special point, $f(I, J) = 1$. Let the value of the spline function at a point x, on a constant-y curve, corresponding to zero mesh point values except for unity at I, be denoted by $F(I, x)$. Then on the dashed line of Figure 5.7, the value of the spline function at all nodes except one vanishes, so that $\phi(x, y) = F(I, x) \cdot G(J, y)$, and so forth. Here G has the same meaning for the y-direction that F has for the x-direction.]

5.10 Given five equally spaced points along the x-axis, construct a spline curve $\phi(x)$ that vanishes at all mesh points except one (consider the three essentially different possibilities). Use the $\phi'' = 0$ end condition, and let $\phi = 1$ at the chosen point. Sketch the resulting curves. How can these results now be used to give an almost immediate spline curve for the general case?

5.11 Carry out the details of the Chebyshev economization example for Equation (5.15), and then carry it one step further to eliminate the term in x^5. What is the new error bound? (The book edited by Abramowitz and Stegun, 1964, gives a table of coefficients for the expression of various powers of x in terms of the T_n, although this is hardly needed here.)

5.12 Observing that

$$T'_n(x) = \frac{dT_n/d\theta}{dx/d\theta} = \frac{d(\cos n\theta)/d\theta}{d(\cos \theta)/d\theta} ,$$

deduce that $T_n(x)$ satisfies the differential equation

$$(1 - x^2)T''_n(x) - xT'_n(x) + n^2 T_n(x) = 0 .$$

5.13 In Equation (5.19) take $a = 0$, $b = 1$ and show that the least squares formulation leads to the coefficient matrix

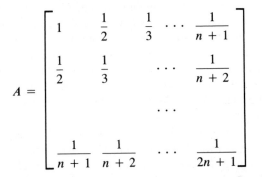

$$A = \begin{bmatrix} 1 & \dfrac{1}{2} & \dfrac{1}{3} & \cdots & \dfrac{1}{n+1} \\ \dfrac{1}{2} & \dfrac{1}{3} & & \cdots & \dfrac{1}{n+2} \\ & & \cdots & & \\ \dfrac{1}{n+1} & \dfrac{1}{n+2} & & \cdots & \dfrac{1}{2n+1} \end{bmatrix}$$

Evaluate the determinant of A for $n = 5$, 10, and 20, and you will see why this *Hilbert matrix* is notorious in numerical analysis. (Ill-conditioning can be expected because the rows are very similar.)

5.14 Let x_k, for $k = 1, 2, \ldots , N$, denote the zeros of $T_N(x)$. Show that, if $m \neq n$,

$$\sum_{k=1}^{N} T_m(x_k) \cdot T_n(x_k) = 0 ,$$

so that the Chebyshev polynomials are orthogonal over this kind of summation. [*Hint:* Observe that $\cos(m + n)\theta = \mathrm{Re}\{\exp[i(m + n)\theta]\}$, etc.] Are there any restrictions on m, n, and N?

5.15 Repeat Problem 5.1, using as interpolation points the (scaled) zeros of the appropriate Chebyshev polynomial, and observe that the fit to the Runge function now improves as N grows.

5.16 Let (x_1, x_2, x_3) be three given x-values, and let (y_1, y_2, y_3, y_4) be four given y-values. Suppose that values of a function $f(x, y)$ are known at the 12 grid points (x_i, y_j), for $i = 1, 2, 3$ and $j = 1, 2, 3, 4$. Then the polynomial

$$L(x, y) = \frac{(x - x_1)(x - x_3)(y - y_1)(y - y_2)(y - y_4)}{(x_2 - x_1)(x_2 - x_3)(y_3 - y_1)(y_3 - y_2)(y_3 - y_4)}$$

vanishes at all grid points other than at (x_2, y_3), where it has unit value. Use this idea to obtain an interpolation polynomial $P(x, y)$ that agrees with $f(x, y)$ at the 12 given points. What is the form of $P(x, y)$ if it is expanded in powers of x and y and like terms are collected? How many coefficients does this resulting polynomial have?

Observe that $P(x, y)$ can be written as a polynomial in x, where the coefficients are themselves various polynomials in y. If a subroutine is available that performs one-dimensional interpolation, then its use in the x-direction gives the values of these coefficient polynomials at each value y_i, so that a new application of the subroutine in the y-direction provides the desired two-dimensional interpolation. Thus, polynomial interpolation in two (or more) dimensions can be obtained by repeated use of a one-dimensional interpolation subroutine.

5.17 Another criterion for the minimization of the absolute value of the factor $\phi_n(x) = (x - x_1)(x - x_2) \cdots (x - x_n)$, over $(-1, 1)$, say, in the error term of Equation (5.17), would be to use the mean square sense; that is, choose the x_i so that $\int_{-1}^{1} \phi_n^2 \, dx$ is a minimum. Show that this leads to a choice of the x_i as the zeros of the Legendre polynomial $P_n(x)$. Show also that the same least squares approach, but now using the weighting factor $(1 - x^2)^{-1/2}$ (which emphasizes contributions near the ends of the interval) again leads to the choice of the zeros of the Chebyshev polynomials for the x_i.

5.18 Let ϕ_m, a polynomial of degree m, be a member of the family of orthogonal polynomials (Sec. 5.5) satisfying $\int_a^b p(x)\phi_m(x)\phi_n(x) \, dx = 0$ for $m \neq n$, where $p(x) > 0$ in (a, b). Show that all zeros of ϕ_m are real, simple, and lie in (a, b). [*Hint:* Suppose there were only k zeros, $k < m$, in this interval; then form

$$I = \int_a^b p(x) \cdot (x - x_1)(x - x_2) \cdots (x - x_k)\phi_m(x) \, dx \ .$$

Observe that the integrand has the same sign throughout (a, b); why does this feature imply a contradiction? Suppose next that one of the zeros, say x_2, were of order higher than 1; then consider

$$\int_a^b \frac{p(x)\phi_m^2(x)}{(x - x_2)^2} \, dx \ . \Big]$$

5.19 Let the polynomials ϕ_n of Problem 5.18 be normalized, so that $\int_a^b p(x)\phi_n^2 \, dx = 1$. Denote the coefficient of x^n in ϕ_n by k_n. Show that a recursion relation of the form

$$\phi_{n+1} = \left(\frac{k_{n+1}}{k_n} x + A_n\right) \phi_n - \frac{k_{n+1}k_{n-1}}{k_n^2} \phi_{n-1}$$

must hold for $n = 1, 2, \ldots$, where A_n is a set of constants. [*Hint:* $\phi_{n+1} - (k_{n+1}/$

$k_n)x\phi_n$ is a polynomial of degree n, and so is expressible in terms of ϕ_0, ϕ_1, ... , ϕ_n. Then use orthogonality.]

5.20 Suppose that values y_i of a function $f(x)$ are known at a set of mesh points x_i. It is sometimes useful to think of x as a function $g(y)$ and to obtain a polynomial interpolation for $g(y)$. Write out a Lagrangian interpolation formula to do this, using the x_i and y_i values. Explain how this *inverse interpolation* idea could be used to iterate towards a zero of $f(x)$; compare with the methods of Chapter 1.

5.21 In Equation 5.2 denote the factor multiplying the term y_i by $L_i(x)$. Show that the polynomial

$$U_i(x) = L_i^2(x) \cdot [1 - 2L_i'(x_i) \cdot (x - x_i)]$$

[where $L'(x)$ denotes the derivative of $L(x)$] vanishes at all x_j points for all values of j other than $j = i$, at which point it equals unity. Show also that $U_i'(x)$ vanishes at all points x_j. Examine similarly the polynomial

$$V_i(x) = (x - x_i)L_i^2(x)$$

and show that

$$P(x) = y_1U_1(x) + y_2U_2(x) + \cdots + y_nU_n(x) + y_1'V_1(x) + \cdots + y_n'V_n(x)$$

gives a polynomial of order $2n - 1$ that takes the value y_i and has slope y_i' at each point x_i. This generalization of Lagrange's formula is called *Hermite's formula*. The error expression of Equation 5.17 now has on its righthand side the term

$$(x - x_1)^2(x - x_2)^2 \cdots (x - x_n)^2 \frac{f^{(2n)}(\xi)}{(2n)!} .$$

Why does this result follow directly from Equation (5.17)? [*Hint:* Think of a total of $2n$ interpolation points consisting of n pairs of slightly separated points.]

5.22 A *rational function* is the ratio of two polynomials. Approximation of a given function $f(x)$ by a rational function $P(x)/Q(x)$, where P and Q are polynomials, can sometimes give much better agreement than approximation by a polynomial; a simple example is $f(x) = 1/x$ near the origin. Consulting the literature as necessary (Chapter 23 of Scheid, 1968 is a good starting point), outline a practical method for rational approximation and give an example.

6

ORDINARY DIFFERENTIAL EQUATIONS

6.1 REDUCTION TO STANDARD FORM

Ordinary differential equations are equations in which ordinary (rather than partial) derivatives of an unknown function appear. If y is a function of x, and if we denote dy/dx by y', d^2y/dx^2 by y'', and so forth, then some examples of ordinary differential equations are

$$y' + 3y = \sin x ,$$

$$y'' + 3xy' - 2y = e^x + x^2 - 1 ,$$

$$yy'' - \sqrt{x^2 + (y')^2} = y'''/(x^2 + 1) . \tag{6.1}$$

The *order* of a differential equation is the order of the highest derivative in it; thus the above equations are of first, second, and third order, respectively. If the unknown function y and its derivatives occur linearly, as in the first two equations, then the equation is *linear*; otherwise (as in the third equation) it is *nonlinear*. The most general form of an nth-order linear differential equation is

$$a_0(x) + a_1(x)y' + a_2(x)y'' + \cdots + a_n(x)y^{(n)} = f(x) , \tag{6.2}$$

where $a_0, a_1, \ldots, a_n, f$ are prescribed functions of x.

Experience and theory show that the solution to a differential equation cannot be determined unless additional data, beyond the equation itself, is given. This data might take the form of a prescription of y and some of its derivatives at one end of the interval of interest, in which case we have an *initial value problem*. Alternatively, some of the data might be given at each end of the interval, and we then have a *boundary value problem.* For example, we might want to solve the second Equation of (6.1) in the interval $(0, 1)$ subject to the condition $y(0) = 1$, $y'(0) = 2$, and this would then be an initial value problem. If instead we were told that $y(0) = -1$, $y(1)$, $= 3$, we would be concerned with a boundary value problem. In some situations, especially when dealing with nonlinear differential equations, the solution may not be uniquely determined by the initial or boundary conditions, so that further information would be required.

Another complication that can arise is that more than one dependent variable may be involved. If y and z are functions of x, then an example of a coupled (nonlinear) differential equation set is given by

$$y'' + xyz + z' = x^2,$$

$$(z')^2 + x(y' + z') + z'' = 0 . \tag{6.3}$$

In general, it is possible to replace single equations of order higher than one, or a set of coupled equations, by a set of first-order equations. Two examples will illustrate the process and make the idea clear.

In the third of Equations (6.1), define three new dependent variables y_1, y_2, y_3 by $y_1 = y$, $y_2 = y'$, $y_3 = y''$. Then $y_3' = y'''$, so that the third equation of (6.1) provides an expression for y_3' in terms of x, y_1, y_2, and y_3. Putting everything together, the result is

$$y_1' = y_2 ,$$

$$y_2' = y_3 ,$$

$$y_3' = (x^2 + 1) [y_1 y_3 - \sqrt{x^2 + y_2^2}] .$$

This is a set of three first-order equations. Back substitution would recover the original equation, so the new set is completely equivalent to the originial equation.

Turning now to Equations (6.3), define $y_1 = y$, $y_2 = y'$, $y_3 = z$, $y_4 = z'$. Then

$$y_1' = y_2 ,$$

$$y_2' = x^2 - xy_1 y_3 - y_4 ,$$

$$y_3' = y_4 ,$$

$$y_4' = -(y_4)^2 - x(y_2 + y_4) ,$$

which is again a set of first-order equations.

In both of these examples the algebra was easy. In more complicated cases we may have to be content with implicit rather than explicit expressions for the various first-order derivatives. However, the possibility of a formal reduction of a differential equation system to an equivalent set of first-order equations means that computer programs for the solution of differential equation sets can be directed towards the general form

$$y_1' = f_1(x, y_1, y_2, \ldots, y_n) ,$$

$$y_2' = f_2(x, y_1, y_2, \ldots, y_n) , \qquad\qquad (6.4)$$

$$\vdots$$

$$y_n' = f_n(x, y_1, y_2, \ldots, y_n) ,$$

where the f_i are functions of the indicated variables, and where $y_1, y_2, \ldots, y_n$ are the n dependent variables.

6.2 TAYLOR SERIES METHODS

In view of Section 6.1 it is possible to describe a numerical method in terms of a single first-order equation; the method can then be easily extended to a more general differential equation set when written in the form of Equations (6.4).

Consider the initial value problem of determining a function $y(x)$ over an interval (a, b) such that

$$y' = f(x, y), \qquad y(a) = y_0 , \quad \text{prescribed.} \qquad (6.5)$$

It follows from differential equation theory that if f and $\partial f/\partial y$ are continuous in an appropriate region of the (x, y) plane, then a unique solution to the problem exists. In seeking a numerical solution to Equation (6.5), we first divide the interval (a, b) into a number n of equal subintervals each of length $h = (b - a)/n$, and then we seek to find an approximation to $y(x)$ at each of these mesh points. Let $x_j = a + jh$, and let y_j denote the approximation to the true solution $y(x_j)$ at x_j. From Taylor's Theorem (Appendix A), we know that the true solution satisfies, for each value of j,

$$y(x_{j+1}) = y(x_j) + hy'(x_j) + e , \qquad (6.6)$$

where the error term e can be written as $\frac{1}{2} h^2 y''(\xi)$ [ξ being some point in (x_j, x_{j+1})], provided y'' is continuous.

Since the error term e can be made as small as desired by choosing h sufficiently small, a natural choice for an equation relating y_j to y_{j+1} is obtained by discarding e in Equation (6.6) and by writing

$$y_{j+1} = y_j + hf(x_j, y_j) , \qquad (6.7)$$

where the value of y' has been replaced by $f(x, y)$ from Equation (6.5). Of course, Equation (6.7) may be rewritten as

$$\frac{y_{j+1} - y_j}{h} = f(x_j, y_j) ,$$

which reflects more directly the approximation to the derivative at the point x_j. Equation (6.7) is often referred to as *Euler's method*. Its use is straightforward, since Equation (6.7) determines y_1, y_2, ... in sequence once y_0 is known.

To illustrate the use of Equation (6.7), let $f(x, y) = x^2 - y^2$, $(a, b) = (0, 1)$, and $y_0 = 1$. Equation (6.5) becomes

$$y' = x^2 - y^2 \text{ for } x \text{ in } (0, 1) , \qquad y(0) = 1 , \qquad (6.8)$$

and Euler's method becomes

$$y_0 = 1 ,$$

$$y_{j+1} = y_j + h(x_j^2 - y_j^2) \qquad \text{for } j = 0, 1, 2, \ldots . \qquad (6.9)$$

If we divide the interval $(0, 1)$ into n subintervals, with $h = 1/n$, and carry out the process (6.9) to determine y_n [which should approximate $y(1)$], we obtain the results given in the first two columns of Table 6.1 for various values of n. In the first column of Table 6.1, a computer carrying eight significant figures was used; in the second column the computer carried 16 significant figures. For 16 significant figures the value approaches the exact value (.7500 1570) more and more closely as n grows; for eight figures, however, the results first approach the exact value and then start moving away from it again. The reason for this is to be found in roundoff error. Each time Equation (6.9) is applied, there is some roundoff error on the part of the computer, and as the number of operations becomes large,

Table 6.1. Computed values of $y(1)$ for $y' = x^2 - y^2$, $y(0) = 1$

	Equation (6.9)		Equation (6.13)	
n	8 significant figures	16 significant figures	8 significant figures	16 significant figures
10	.7107 9124	.7107 9126	.7504 1332	.7504 1336
20	.7309 1225	.7309 1229	.7501 0499	.7501 0510
40	.7405 8644	.7405 8652	.7500 3679	.7500 3695
80	.7453 3101	.7453 3118	.7500 2056	.7500 2088
160	.7476 8057	.7476 8088	.7500 1633	.7500 1698
320	.7488 4952	.7488 5014	.7500 1473	.7500 1602
640	.7494 3209	.7494 3338	.7500 1326	.7500 1578
1280	.7497 2215	.7497 2466	.7500 1091	.7500 1572
2560	.7498 6523	.7498 7020	.7500 0501	.7500 1571
5120	.7499 3288	.7499 4296	.7499 9551	.7500 1570
10,240	.7499 5908	.7499 7933	.7499 6226	.7500 1570
20,480	.7499 5960	.7499 9752	.7499 5960	.7500 1570
40,960	.7499 2532	.7500 0661	.7499 2532	.7500 1570
81,920	.7498 4868	.7500 1116	.7498 4868	.7500 1570

the accumulated error can be substantial. The particular computer used is one that rounds *down* by dropping all figures after the last. A computer that rounds either up or down, depending on the size of the next digit, behaves somewhat better as far as roundoff is concerned, but even here the accumulated error can be expected to grow (in proportion perhaps to $\sqrt{n}$; see Chap. 4).

This first example indicates that in order to reduce the effect of roundoff error and to achieve an accurate result with a reasonable number of steps, one should seek a more accurate procedure than that given by Equation (6.7). An obvious possibility is to use two terms of a Taylor formula, so that the exact solution is written as

$$y(x_{j+1}) = y(x_j) + hy'(x_j) + \tfrac{1}{2} h^2 y''(x_j) + e , \qquad (6.10)$$

where the error term is now given by $e = \tfrac{1}{6} h^3 y'''(\xi)$, ξ being again some point in the interval (x_j, x_{j+1}). We know that $y' = f(x, y)$, so that

$$y'' = \frac{d}{dx} f(x, y)$$

$$= \frac{\partial f}{\partial x} + \frac{\partial f}{\partial y} y' \qquad \text{(by the usual chain rule)}$$

$$= f_x + f_y f . \qquad (6.11)$$

Here a subscript denotes a partial derivative. The corresponding mesh value equation is

$$y_{j+1} = y_j + hf + \tfrac{1}{2} h^2 (f_x + f_y f) , \qquad (6.12)$$

where all of f, f_x, f_y are evaluated at (x_j, y_j). For the example problem of Equation (6.8), we have

$$y_{j+1} = y_j + h(x_j^2 - y_j^2) + \tfrac{1}{2} h^2 [2x_j + (-2y_j)(x_j^2 - y_j^2)] , \qquad (6.13)$$

with $y_0 = 1$, as before. The second two columns of Table 6.1 give the results obtained from Equation (6.13). The effects of the higher accuracy formula and of the increased roundoff error (there are more operations in each step) are both evident.

Comparing Equations (6.6) and (6.7), we see that the true solution $y(x)$ fails to satisfy the numerical equation (6.7) by the amount $e = \tfrac{1}{2} h^2 y''(\xi)$, called the *truncation error* or *discretization error*. Because of the factor h^2, we say the truncation error is of order h^2, or of second order; Equation (6.7) itself is described as being of *first-order accuracy*. Similary, Equation (6.12) is of second-order accuracy, the truncation error $\tfrac{1}{6} h^3 y'''(\xi)$ being of order h^3. If roundoff error is not important, then the error made in each application of Equation (6.7) is $\tfrac{1}{2} h^2 y''(\xi)$, where ξ will have a different value each time. If y'' does not change too much over the interval, we could say that the average error for each application of Equation

(6.7) is $\frac{1}{2} h^2(y'')_{av}$, so that in n applications the accumulated error should be $n[\frac{1}{2} h^2(y'')_{av}] = \frac{1}{2} hy''_{av}(b - a)$, since $nh = b - a$. In the case of the example problem, suppose we first use a value h_1 for h and obtain a result Y_1 at $x = 1$. Repeat the process with a value $h_2 = h_1/2$, so as to obtain Y_2 at $x = 1$. Then the true solution must differ from each Y_1 and Y_2 by the appropriate accumulated error, so (using $b - a = 1$)

$$y(1) \cong Y_1 + \frac{1}{2} h_1 y''_{av} , \qquad y(1) \cong Y_2 + \frac{1}{2}(\frac{1}{2} h_1) y''_{av} , \qquad (6.14)$$

where we have used $h_2 = \frac{1}{2} h_1$. Elimination of the term $h_1 y''_{av}$ from Equations (6.14) gives

$$y(1) \cong 2Y_2 - Y_1 . \qquad (6.15)$$

This is called *Richardson extrapolation*. To see how well it works, let us use the $n = 160$ and $n = 320$ results from column 2 of Table 6.1; Equation (6.15) gives

$$y(1) \cong 2(.7488\ 5014) - .7476\ 8088$$

$$\cong .7500\ 1940 ,$$

which is rather good.

For the case of Equation (6.12), Equations (6.14) would be replaced by

$$y(1) \cong Y_1 + \frac{1}{6} h_1^2 y'''_{av} , \qquad y(1) \cong Y_2 + \frac{1}{6}(\frac{1}{2} h_1)^2 y'''_{av}$$

(where $nh = 1$ as before), so that Equation (6.15) is replaced by

$$3y(1) \cong 4Y_2 - Y_1 .$$

With $n = 160$ and $n = 320$, the fourth column of Table 6.1 now gives $y(1) \cong$ .7500 1570, which is correct to eight figures. Richardson extrapolation is less useful when the results are contaminated by roundoff error.

Third- and higher-order Taylor series methods can be obtained similarly; Problem 6.1 describes a third-order method.

We have not actually proved that a process like that of Equations (6.7) or (6.12) will yield results that (in the absence of roundoff error) converge to the true solution as $h \to 0$. For appropriate functions $f(x, y)$, a formal proof of this fact can be given. However, we content ourselves here with a consideration of the special case $f(x, y) = \lambda y$, where λ is some constant. This is a standard trial example in numerical analysis; not only is it useful (as we will see) in stability investigations, it also provides an equation prossessing an exact solution in simple form for testing the accuracy of algorithms.

With $f(x, y) = \lambda y$, Equation (6.7) becomes

$$y_{j+1} = y_j + \lambda h y_j = (1 + \lambda h) y_j ,$$

so that with $y(a) = y_0$ given, we find

$$y_1 = y_0(1 + \lambda h) , \qquad y_2 = y_0(1 + \lambda h)^2 ,$$

and, for any choice of j,

$$y_j = y_0(1 + \lambda h)^j .$$

Choose a fixed point x in (a, b), and let there be N intervals of length h between a and x. Then the numerical solution is

$$y_N = y_0(1 + \lambda h)^N$$

$$= y_0\left(1 + \lambda \frac{x - a}{N}\right)^N$$

$$= y_0 \exp\left[N \ln\left(1 + \lambda \frac{x - a}{N}\right)\right]$$

$$= y_0 \exp\left[N\left\{\lambda \frac{x - a}{N} - \frac{1}{2}\left(\lambda \frac{x - a}{N}\right)^2 + \frac{1}{3}\left(\lambda \frac{x - a}{N}\right)^3 - \cdots\right\}\right],$$

where the series expansion for the ln function has been used. We can choose N large enough so that the error caused by dropping all terms other than the first is smaller than any desired quantity; therefore, as $N \to \infty$,

$$y_N \to y_0 \exp\left[N\left(\lambda \frac{x - a}{N}\right)\right] \to y_0 \exp[\lambda (x - a)] ,$$

which is, in fact, the exact solution of $y' = \lambda y$, $y(a) = y_0$. This completes the convergence proof for this example.

A pictorial representation of Equation (6.7) is given in Figure 6.1. It is clear that even if y_j happens to agree exactly with the true solution, $y(x_j)$, the value of y_{j+1} will differ from $y(x_{j+1})$ because of the use of the tangent at x_j rather than of some average tangent. This suggests that a more accurate algorithm would be

$$y_{j+1} = y_j + \tfrac{1}{2} h[f(x_j, y_j) + f(x_{j+1}, y_{j+1})] , \qquad (6.16)$$

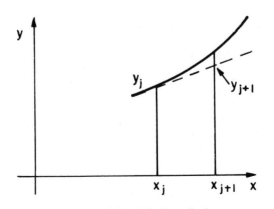

Figure 6.1. Euler's method.

in which we use the average of the slopes at the two points (x_j, y_j) and (x_{j+1}, y_{j+1}). Unfortunately, we do not know y_{j+1} for use in the righthand side of Equation (6.16), so that we cannot use Equation (6.16) to determine y_{j+1} *explicitly*, as we could in Equation (6.7). The value of y_{j+1} is now determined *implicitly* by Equation (6.16) itself. For the example problem in which $f = x^2 - y^2$, we must solve

$$y_{j+1} = y_j + \frac{h}{2}[(x_j^2 - y_j^2) + (x_{j+1}^2 - y_{j+1}^2)]$$

at each step for y_{j+1}. Here this is not difficult because the equation is only a quadratic (although we must choose the right root each time), but in general we would have to use one of the iterative nonlinear equation solvers of Chapter 1. (The previous y_j value provides a good starting guess for y_{j+1}.) Thus we pay for whatever increased accuracy Equation (6.16) provides by having to work harder at each step. Implicit equations do, however, often give a bonus, in that the kind of "feedback" associated with the appearance of y_{j+1} on the righthand side can enhance the stability of the algorithm.

It is a useful exercise to determine the accuracy of Equation (6.16). Again we inquire as to the extent by which the true solution fails to satisfy Equation (6.16). For the true solution we have (from a Taylor expansion)

$$y(x_{j+1}) = y(x_j) + hy'(x_j) + \tfrac{1}{2}h^2 y''(x_j) + \tfrac{1}{6}h^3 y'''(x_j) + \cdots . \quad \textbf{(6.17)}$$

But this is not quite what we get from the righthand side of Equation (6.16), which for the true solution equals (with $f = y'$, etc.)

$$y_j + \tfrac{1}{2} h[y_j' + (y_j' + hy_j'' + \tfrac{1}{2}h^2 y_j''' + \cdots)] ,$$

where the quantity in parentheses is obtained from a Taylor expansion of $f(x_{j+1}, y_{j+1})$, that is y_{j+1}', and where the subscript j means evaluation at j. We see that there is a discrepancy in the terms of order h^3, so that we could write Equation (6.16) more indicatively as

$$y_{j+1} = y_j + \tfrac{1}{2} h[f(x_j, y_j) + f(x_{j+1}, y_{j+1})] - \tfrac{1}{12} h^3 y'''(\xi) .$$

Thus, the implicit algorithm of Equation (6.16) is of second-order accuracy and has a truncation error of order h^3. A somewhat different implicit method is given in Problem 6.3.

These methods are easily generalized to apply to sets of equations. To apply the Euler method to the equation pair $y' = f(x, y, z)$, $z' = g(x, y, z)$, where y and z are functions of the independent variable x, we use

$$y_{j+1} = y_j + hf(x_j, y_j, z_j) ,$$

$$z_{j+1} = z_j + hg(x_j, y_j, z_j) .$$

For the second-order Taylor algorithm, we would use

$$y_{j+1} = y_j + hf + \tfrac{1}{2} h^2 [f_x + f_y f + f_z g] \; ,$$

$$z_{j+1} = z_j + hg + \tfrac{1}{2} h^2 [g_x + g_y f + g_z g] \; ,$$

where subscripts denote partial derivatives and all quantities on the righthand sides of these equations are to be evaluated at (x_j, y_j, z_j). Sets of more than two equations are treated similarly.

6.3 RUNGE–KUTTA METHODS

In Equation (6.10) one more term, namely, $\tfrac{1}{6} h^3 y'''(x_i)$, could have been included, which would have corresponded to the addition of the term

$$\tfrac{1}{6} h^3 (f_{xx} + 2 f_{xy} f + f_{yy} f^2 + f_x f_y + f_y^2 f)$$

to the righthand side of Equation (6.12), which would now be of third-order accuracy. This process can be extended to give a Taylor series method of any desired order of accuracy. Its drawback is that the various partial derivatives of f can be complicated expressions, time-consuming to compute. It is desirable to have available a high-order method that requires evaluations of f alone, and this the Runge–Kutta idea provides.

We begin with a second-order Runge–Kutta method in order to keep the algebra fairly simple. Again suppose that we want to solve $y' = f(x, y)$ in an interval (a, b), where $y(a) = y_0$ is prescribed, and that the interval has been divided into n-subintervals each of length $h = (b - a)/n$. Define $x_j = a + jh$ and denote the computed approximation to the true solution $y(x_j)$ at this point by y_j. Then the Runge–Kutta method carries out the step from x_j to x_{j+1} in stages:

$$k_1 = f(x_j, y_j) \; ,$$

$$k_2 = f(x_j + \alpha h, y_j + \beta h k_1) \; ,$$

$$y_{j+1} = y_j + h[A_1 k_1 + A_2 k_2] \; , \tag{6.18}$$

where α, β, A_1, and A_2 are constants to be determined. We must choose these constants so that the amount by which the true solution fails to satisfy Equations (6.18) is as small as possible.

For the true solution $y(x)$, the quantity k_1 would equal the exact slope at the jth point, and by use of a Taylor expansion in two variables (Appendix D) the quantity k_2 would equal

$$
\begin{aligned}
f(x_j + \alpha h, y_j + \beta h f) = {} & f + f_x \alpha h + f_y \beta h f \\
& + \tfrac{1}{2} [f_{xx}(\alpha h)^2 + 2 f_{xy}(\alpha h)(\beta h f) + f_{yy}(\beta h f)^2] \\
& + \tfrac{1}{6} [f_{xxx}(\alpha h)^3 + \cdots] \; ,
\end{aligned}
$$

where subscripts on f indicate partial derivatives and all quantities are evaluated at the point j. In the third equation (6.18), the various powers of h on the righthand side can be collected to give

$$h^0: \quad y_j ,$$

$$h^1: \quad A_1 f_{\cdot} + A_2 f ,$$

$$h^2: \quad A_2(f_x \alpha + f_j \beta f) , \qquad\qquad (6.19)$$

$$h^3: \quad \tfrac{1}{2} A_2 [f_{xx} \alpha^2 + 2 f_{xy} \alpha \beta f + f_{yy} \beta^2 f^2] ,$$

$$\vdots$$

The lefthand side, still for the true solution, would be given by the usual Taylor expansion as

$$y(x_{j+1}) = y(x_j) + hf + \tfrac{1}{2} h^2(f_x + f_y f)$$

$$+ \tfrac{1}{6} h^3 (f_{xx} + 2 f_{xy} f + f_{yy} f^2 + f_y f_x + f_y^2 f) , \qquad (6.20)$$

where we have used $df/dx = f_x + f_y f$, and so forth. Comparison shows that we get agreement (for all functions f) up to terms of order h^2 by the choice of coefficients

$$A_1 + A_2 = 1 , \quad A_2 \alpha = \tfrac{1}{2} , \quad A_2 \beta = \tfrac{1}{2} , \qquad (6.21)$$

but that we cannot, in general, have agreement to higher order (for all f) because of the different ways in which the function f enters into the coefficients of h^3 in Equations (6.19) and (6.20). There is more than one way in which we can satisfy Equations (6.21). One popular choice is $A_1 = A_2 = \tfrac{1}{2}$, $\alpha = \beta = 1$; another is $A_1 = 0$, $A_2 = 1$, $\alpha = \beta = \tfrac{1}{2}$.

The first choice leads to

$$k_1 = f(x_j, y_j) ,$$

$$k_2 = f(x_j + h, y_j + hk_1) ,$$

$$y_{j+1} = y_j + \tfrac{1}{2} h[k_1 + k_2] , \qquad (6.22)$$

and the second leads to

$$k_1 = f(x_j, y_j) ,$$

$$k_2 = f(x_j + \tfrac{1}{2} h, y_j + \tfrac{1}{2} hk_1) ,$$

$$y_{j+1} = y_j + hk_2 . \qquad (6.23)$$

Each of these is of second-order accuracy and does not require evaluations of the derivatives of f.

Runge–Kutta methods of higher order are derived similarly (Prob. 6.5 describes a third-order method). There usually is some freedom in the choice of coefficients, as in the second-order case above, and arguments as to the advantage of one choice compared to another are sometimes made, although it is the author's experience that these advantages are usually minor. As a compromise between order of accuracy and computational labor per step, the most common choice is probably a fourth-order method; one such is

$$k_1 = f(x_j, y_j) \, ,$$

$$k_2 = f(x_j + \tfrac{1}{2}h, y_j + \tfrac{1}{2}hk_1) \, ,$$

$$k_3 = f(x_j + \tfrac{1}{2}h, y_j + \tfrac{1}{2}hk_2) \, ,$$

$$k_4 = f(x_j + h, y_j + hk_3) \, ,$$

$$y_{j+1} = y_j + \frac{h}{6}(k_1 + 2k_2 + 2k_3 + k_4) + O(h^5) \, , \qquad \textbf{(6.24)}$$

where the term $O(h^5)$ means "with an error of order h^5," corresponding to the fourth-order accuracy of the algorithm.

The following subroutine* implements the fourth-order procedure of Equations (6.24) for the case of a set of n first-order equations; the reader should note the simple way in which the single-equation case is extended to the n-equations case.

```
C   SUBROUTINE RUKU(N,NS,X,Y,H,XK,YY) USES A FOURTH ORDER
C   RUNGE-KUTTA METHOD TO SOLVE A SET OF FIRST ORDER EQUATIONS
C   OF THE FORM   DY(I)/DX=FUN(I,X,Y)   , FOR I=1,2,...,N.
C
C   INPUT:
C      N = NUMBER OF EQUATIONS
C      NS = NUMBER OF STEPS TO BE TAKEN
C      X = INITIAL VALUE OF INDEPENDENT VARIABLE
C      Y = ARRAY OF INITIAL VALUES OF THE Y(I)   (DIMENSION N)
C      H = STEP SIZE (NEGATIVE IF X IS TO DECREASE)
C      XK = WORKING ARRAY (DIMENSION (4,N))
C      YY = WORKING ARRAY (DIMENSION N)_
C
C   OUTPUT:
C      X = FINAL VALUE OF X
C      Y = ARRAY OF FINAL Y VALUES
C
C   FUNCTION SUBPROGRAM CALLED:   FUN(I,X,Y)
C
      SUBROUTINE RUKU(N,NS,X,Y,H,XK,YY)
      DIMENSION Y(N),XK(4,N),YY(N)
      H2=H/2.
      H6=H/6.
      DO 5 I=1,NS
      XH2=X+H2
```

---

*An actual library subroutine might differ slightly from the one given here. See Problem 6.7.

```
      XH=X+H
      DO 6 J=1,N
      XK(1,J)=FUN(J,X,Y)
6     YY(J)=Y(J)+H2*XK(1,J)
      DO 7 J=1,N
      XK(2,J)=FUN(J,XH2,YY)
7     YY(J)=Y(J)+H2*XK(2,J)
      DO 8 J=1,N
      XK(3,J)=FUN(J,XH2,YY)
8     YY(J)=Y(J)+H*XK(3,J)
      DO 9 J=1,N
      XK(4,J)=FUN(J,XH,YY)
9     Y(J)=Y(J)+H6*(XK(1,J)+2.*(XK(2,J)+XK(3,J))+XK(4,J))
      X=XH
5     CONTINUE
      END
```

It is also possible to devise implicit Runge–Kutta schemes. A simple one, for the case $y' = f(x, y)$, is

$$k_1 = f(x_j, y_j) ,$$

$$k_2 = f(x_j + \frac{h}{2}, y_j + \frac{h}{4}[k_1 + k_2]) ,$$

$$k_3 = f(x_j + h, y_j + hk_2) ,$$

$$y_{j+1} = y_j + \frac{h}{6}(k_1 + k_2 + k_3) + O(h^5) . \tag{6.25}$$

In this fourth-order method the equation determining k_2 is implicit. The saving of one step in Equations (6.25), as compared to Equations (6.24), is usually more than compensated for by the added labor in solving the implicit equation.

The Runge–Kutta method lends itself well to "adaptive" situations in which one changes the step length as the program proceeds. Every few steps, a higher-order Runge–Kutta method can be used for the next step, and the result compared with that obtained by the regular method; if the results are significantly different, the step size is reduced and the process continues. If the results are almost identical, the step size could similarly be increased.

Whether one plans to have the program itself adjust the step size or not, the importance of error checking in solution algorithms for differential equations can hardly be overemphasized. Any computer results should be considered purely tentative until the effects of a different step size have been examined and until similar problems of known exact solution have been solved. It is also worthwhile to explore the effect of small changes in the problem parameters or in the initial conditions. The parameters h is of particluar importance; it must be small enough that the Runge-Kutta algorithm accurately approximates the differential equations, and it must also be small compared to the distance over which the solution is undergoing significant change. For one restriction on h, see Problem 6.20.

6.4 MULTISTEP METHODS

In solving $y' = f(x, y)$ step by step, values of y_j accumulate, and it is reasonable to try to use these past values (and the values of slope f as determined at these points) to improve the prediction of y_{j+1}. A simple example of this idea is given by

$$y_{j+1} = y_{j-1} + 2hf(x_j, y_j) , \qquad (6.26)$$

which uses information associated with y_j and y_{j-1} in order to obtain y_{j+1}. The usual Taylor expansion shows that the true solution fails to satisfy Equation (6.26) by a term of $O(h^3)$, so that Equation (6.26) has second-order accuracy. A graphical interpretation of Equation (6.26) is given in Figure 6.2, which shows that the derivative to be used in going from y_{j-1} to y_{j+1} is evaluated at the midpoint j, and it is plausible that this is a more accurate procedure than would result from an evaluation of the derivative at an endpoint (either $j - 1$ or $j + 1$). Equation (6.26) is an example of a *multistep method* (as contrasted to the *single-step methods* of Secs. 6.2 and 6.3) because it uses information at more than one past step.

There is an immediate difficulty associated with Equation (6.26): It is that it is not self-starting. Suppose that the mesh points along the x-axis are labeled x_0, $x_1, x_2, \ldots$, and the corresponding y-values are labeled $y_0, y_1, y_2, \ldots$. An initial condition would prescribe y_0; then if y_1 were known, we could apply Equation (6.26) to find y_2, then use y_1 and y_2 to find y_3, and so on. However, y_1 is *not* known, so some other method must be used in order to get Equation (6.26) started. In this case we could choose a Taylor series method or a Runge–Kutta method of at least second-order accuracy to be compatible with the order of accuracy of

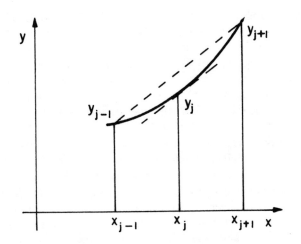

Figure 6.2. Midpoint method.

Equation (6.26). The same difficulty arises if we want to change step size during a run. Suppose we have computed ... , y_{80}, y_{81}, y_{82} with a step size h, and at this point we decide (perhaps after comparison with results obtained by the use of a higher-order algorithm) that the error is becoming unacceptable and the mesh size should be halved. To proceed to the next mesh point by using Equation (6.26) requires knowledge of y_{82} and also of y_r, where x_r lies halfway between x_{81} and x_{82}; to obtain y_r, we could use either a single-step method starting from (x_{81}, y_{81}), or we could interpolate.

Equation (6.26) also suffers from a more subtle difficulty. Suppose we consider the special equation $y' = \lambda y$, where y is a constant , with the initial condition $y(0) = 1$. The exact solution is then $y = e^{\lambda x}$. From Equation (6.26) we obtain

$$y_{j+1} = y_{j-1} + 2h\lambda y_j , \tag{6.27}$$

which is a *difference equation* in the y_j. We now observe that if we try $y_j = r^j$ in Equation (6.27), where r is some constant, then we get a special solution of the equation if the constant r satisfies the quadratic equation

$$r^2 - 2h\lambda r - 1 = 0 .$$

There are two possible values of r:

$$r_1 = h\lambda + \sqrt{1 + h^2\lambda^2} \quad \text{and} \quad r_2 = h\lambda - \sqrt{1 + h^2\lambda^2} .$$

Thus, essentially by guesswork, we have found two special solutions of Equation (6.27); in fact, any linear combination is also a solution. Consequently, a particular solution of Equation (6.27) is given by

$$y_j = Ar_1^j + Br_2^j , \tag{6.28}$$

where A and B are arbitrary constants. Observe next that we can choose A and B so that the righthand side of Equation (6.28) has any desired value for $j = 0$ and any desired value for $j = 1$. But this means that Equation (6.28) is, in fact, the general solution of Equation (6.27), since (6.27) determines all of y_2, y_3, ... explicitly if y_0 and y_1 are given; hence, there can be only one solution satisfying these conditions, and Equation (6.28) must be it. It follows that Equation (6.28) gives the solution of Equation (6.27) that the computer would generate (for given y_0 and y_1) in the absence of roundoff error.

Suppose now that λ is negative. Then the true solution $e^{\lambda x}$ decays as x increases. On the other hand, the righthand side of Equation (6.28) does not decay because $|r_2| > 1$; therefore, except in the fortuitous case $B = 0$, the computed solution would diverge exponentially from the true solution as x becomes large. Although it is true that by a bit of ingenuity we could choose the value of y_1 to make $B = 0$, this device does not protect us against the possibility that roundoff error could inject some of the r_2^j "ingredient" at any stage of the calculation, which would eventually devastate the numerical process. We can expect, then, that Equation

(6.27) is *unstable*, at least if unbounded values of x are to be considered.* More-over, the special problem $y' = \lambda y$ is not as special as it might at first appear, for if we are solving $y' = f(x, y)$ and make a small change in y (perhaps roundoff error again), say $\delta y(x)$, then the equation becomes

$$(y + \delta y)' = f(x, y + \delta y) = f(x, y) + f_y \delta y$$

to the first order, where f_y denotes $\partial f/\partial y$. Subtracting the original solution $y' = f(x, y)$ shows that the "peturbation" δy satisfies $(\delta y)' = f_y(\delta y)$, and if f_y is given its average value over a few steps, we end up again with Equation (6.27), now governing the growth of the perturbation.

There are many multistep methods in the literature. The most commonly used explicit methods are probably the Adams–Bashforth formulas (obtained as before by the use of Taylor expansions):

$$y_{j+1} = y_j + \frac{h}{2} [3f(x_j, y_j) - f(x_{j-1}, y_{j-1})] + \frac{5}{12}h^3 y'''(\xi) ,$$

$$y_{j+1} = y_j + \frac{h}{12} [23f(x_j, y_j) - 16f(x_{j-1}, y_{j-1}) + 5f(x_{j-2}, y_{j-2})] + \frac{3}{8}h^4 y^{(4)}(\xi) ,$$

$$y_{j+1} = y_j + \frac{h}{24} [55f(x_j, y_j) - 59f(x_{j-1}, y_{j-1}) + 37f(x_{j-2}, y_{j-2})$$

$$- 9f(x_{j-3}, y_{j-3})] + \frac{251}{720}h^5 y^{(5)}(\xi) ,$$

where the error term, again, refers to the extent by which the true solution fails to satisfy the equations. Here ξ is some value of x in the interval spanned by the x_j values. For appropriate values of h, these equations do not suffer from the possible instability problem experienced in connection with Equation (6.26).

There is a comparable set of implicit multistep methods, known as the Adams–Moulton formulas:

$$y_{j+1} = y_j + \frac{h}{2} [f(x_{j+1}, y_{j+1}) + f(x_j, y_j)] - \frac{h^3}{12}y'''(\xi) ,$$

$$y_{j+1} = y_j + \frac{h}{12} [5f(x_{j+1}, y_{j+1}) + 8f(x_j, y_j) - f(x_{j-1}, y_{j-1})] - \frac{h^4}{24}y^{(4)}(\xi) ,$$

---

*If the range of x is fixed, Equation (6.27) can still give good results as the step size h is decreased. For that reason the instability associated with Equation (6.27) is termed *weak instability*.

$$y_{j+1} = y_j + \frac{h}{24} [9f(x_{j+1}, y_{j+1}) + 19f(x_j, y_j) - 5f(x_{j-1}, y_{j-1})$$

$$+ f(x_{j-2}, y_{j-2})] - \frac{19}{720}h^5 y^{(5)}(\xi) ,$$

Note that the coefficients of the error terms in the implicit formulas are smaller than the corresponding ones in the explicit formulas. This is not surprising, in view of the fact that the implicit formulas evaluate derivatives closer to the scene of the action. This feature is sometimes taken advantage of by implementing a *predictor–corrector* algorithm. The idea is to use a fourth-order Adams–Bashforth method, for example, to "predict" the value of y_{j+1} and to then use this value as a starting value for y_{j+1} in an iterative solution of the fourth-order Adams–Moulton formula. One would not want to use more than one or two repeated substitutions, however, in order to avoid excessive labor.

An advantage of the predictor–corrector method is that an estimate for the error can be made. For example, if the fourth-order algorithms of the Adams–Bashforth and Adams–Moulton formulas are used, with $\bar{y}_{j+1}$ denoting the predictor result, y_{j+1} the corrector result, and Y_{j+1} the exact solution, we have

$$Y_{j+1} \cong \bar{y}_{j+1} + \frac{251}{720}h^5 y^{(5)}(\xi_1) ,$$

$$Y_{j+1} \cong y_{j+1} - \frac{19}{720}h^5 y^{(5)}(\xi_2) ,$$

where ξ_1 and ξ_2 are not necessarily equal. If h is small, and if $y^{(5)}$ does not change too much over the interval containing the mesh points used in these two formulas, it is reasonable to assume that the terms $h^5 y^{(5)}$ are approximately equal in the two equations. Subtraction of the second equation from the first determines $h^5 y^{(5)}$, and an estimate for the error after the use of the corrector formula is then given by

$$-\frac{19}{720}h^5 y^{(5)} \cong -\frac{19}{270}(y_{j+1} - \bar{y}_{j+1}) .$$

This quantity could be added to the y_{j+1} value given by the corrector formula, but it is perhaps better to keep h small enough so that this estimated error is acceptable. This kind of error estimate is sometimes termed *Milne's device*. Incidentally, there is a well-known predictor–corrector method associated with the name Milne; see Problem 6.9.

Returning to multistep methods in general, a possible advantage, as compared to a Runge–Kutta method, is that the number of functional evaluations is reduced. Thus, if the fourth-order Adams–Bashforth method is used for the equation $y' = f(x, y)$, it requires only one new evaluation of $f(x, y)$ to take a new step; the fourth-order Runge–Kutta method of Equations (6.24) requires three new evaluations. If

$f(x, y)$ is complicated, this consideration may outweigh the starting and step-change difficulties associated with a multistep method. Of course, if a predictor–corrector method is used, the added problem of iterating the predictor to convergence must be considered. Moreover, the fact that a multistep method replaces a differential equation with a higher-order difference equation can result in stability problems, as noted in connection with Equation (6.26) (and see also Prob. 6.21).

6.5 BOUNDARY VALUE PROBLEMS

It usually only makes sense to have data prescribed at the two ends of an interval when the differential equation system is of at least second order. For example, let

$$y'' + xy' + 2y = x^2 \tag{6.29}$$

for $0 < x < 1$, with $y(0) = 1$, $y(1) = 2$. The equation may be rewritten as an equation pair by the usual device. Set $y_1 = y$, $y_2 = y'$, to obtain

$$y_1' = y_2 , \qquad y_2' = x^2 - xy_2 - 2y_1 . \tag{6.30}$$

We have $y_1(0) = 1$, but we don't know $y_2(0)$. In effect, we have to choose $y_2(0)$ in such a way that $y_1(1)$ will turn out to be 2.

The linearity of Equation (6.29) makes this easy. First, suppose that we obtain (numerically) a solution $w(x)$ to Equation (6.29) satisfying the initial conditions $w(0) = 1$, $w'(0) = 0$. Second, obtain a different solution $z(x)$ to Equation (6.29) satisfying $z(0) = 1$, $z'(0) = 1$. Then the linear combination

$$y(x) = Aw(x) + Bz(x) , \tag{6.31}$$

where A and B are constants, will satisfy both Equation (6.29) and the condition $y(0) = 1$, provided $A + B = 1$. Having recorded $w(1)$ and $z(1)$, we now require $Aw(1) + Bz(1) = 2$; these two equations determine A and B, and Equation (6.31) then gives the desired solution. (We find, in fact, $w(1) = .344093$, $z(1) = .950531$, so $B = 2.730546$, which means $y'(0) = B = 2.730546$. As a check, the numerical solution with this value for $y'(0)$ did indeed give $y(1) = 2$.)

As a second example, alter Equation (6.29) to read

$$y'' + xy' + 2y^2 = x^2 \tag{6.32}$$

(i.e., the quantity y has been replaced by y^2). Consider the boundary conditions $y(0) = 1$, $y(1) = 1.5$. The equation is now nonlinear, so the previous superposition process is inapplicable. What we can do, however, is guess $y'(0)$ and find the corresponding $y(1)$; the guess for $y'(0)$ can now be repeatedly improved until $y(1) = 1.5$. We think of $y(1)$ as being a nonlinear function of $y'(0)$, so the methods of Chapter 1 are applicable. The completion of this problem is left as an exercise for the reader. Of course, there is, in general, no guarantee that a nonlinear boundary value problem will have a unique solution or, for that matter, any solution at all.

The method described above, in which we try to find $y'(0)$ by intelligent experimentation, is called a *shooting method*, for obvious reasons. Another approach to the solution of a boundary value problem is to replace it by a set of coupled algebraic equations. Consider again the problem of Equation (6.29) (although the method only comes into its own for harder problems) and divide the interval (0, 1) into n subintervals of length h. Denote the value of y at the jth mesh point ($j = 0, 1, \cdots, n$) by y_j. Although we could use a coupled pair of first-order equations, let us work directly with Equation (6.29) just for variety. By a Taylor expansion the reader may verify that approximations to y' and y'' at the point j are given by

$$y' \cong \frac{y_{j+1} - y_{j-1}}{2h}, \qquad y'' = \frac{y_{j+1} - 2y_j + y_{j-1}}{h^2},$$

so that Equation (6.29) can be approximated by

$$\frac{y_{j+1} - 2y_j + y_{j-1}}{h^2} + x_j \frac{y_{j+1} - y_{j-1}}{2h} + 2y_j = x_j . \qquad \textbf{(6.33)}$$

Now y_0 and y_n are known from the given data, so that Equations (6.33), written for $j = 1, 2, \ldots, n - 1$, are a set of $n - 1$ equations in $n - 1$ unknowns. The coefficient matrix is tridiagonal (see Prob. 2.13), so their solution is almost immediate—one forward sweep and one backward sweep do it.

A similar process can be applied to the nonlinear equation (6.32). Equations (6.33) are now replaced by a set of nonlinear algebraic equations, each involving only three adjacent y_j values. The methods of Chapter 3 are applicable here.

There are some problems that are very difficult to solve by a shooting method but for which the algebraic equation method is very effective. Among these are *boundary layer problems*, so-called because of the fact that many of them are encountered in fluid mechanics problems involving boundary layers (a boundary layer is a fluid region near a wall where the velocity changes rapidly from its free-stream value to zero at the wall). However, problems of similar type arise in many other fields. As an example of a problem of this kind, consider

$$\epsilon y'' - (3 - x^2)y = -1 \qquad \textbf{(6.34)}$$

for $-1 < x < 1$, with $y(-1) = 0$, $y(1) = 0$. Here ϵ is a constant, much less than unity in magnitude (a typical physical value is $\epsilon = 10^{-6}$). To see that the shooting method would fail here, replace Equation (6.34) temporarily by an equation in which $3 - x^2$ is given an "average" constant value, say k. Then the exact solution to $\epsilon y'' - ky = -1$ with $y(-1) = 0$, $y'(-1) = \alpha$ [where α is our guess for $y'(-1)$] is easily found, and it is of order $\exp(2\sqrt{k/\epsilon})$ at $x = 1$. For $k \cong 2$, $\epsilon \cong 10^{-6}$, $\alpha \cong 1$, this quantity is of order 10^{1000}, which is out of reasonable computer range. The algebraic equation method, however, will work well on this problem, particularly if one uses a variable mesh spacing that becomes small near the ends. Some nonlinear boundary value problems are also readily treated by the discretization method. If ϵ is very small, it is usually advisable to approach the final mesh iteratively by solving a sequence of problems for successively smaller values of ϵ. For each new choice of ϵ the previous mesh is tried first; if y is found to change too much between mesh points, then new intermediate mesh points are inserted, and the problem is solved again.

6.6 STIFF PROBLEMS

It may be verified by substitution that the exact solution of

$$w'' + 1001w' + 1000w = 0 \qquad w(0) = 1 , \quad w'(0) = -1 , \qquad \textbf{(6.35)}$$

is given by $w = e^{-x}$. This is a well-behaved function in, say, $0 < x < 1$, with derivatives of modest size. Presumably a Runge–Kutta method would work well, even for step sizes as large as perhaps .1. If Equation (6.35) is written as an equation pair (with $y = w$, $z = w'$)

$$y' = z , \qquad z' = -1001z - 1000y ,$$
$$y(0) = 1 , \qquad z(0) = -1 , \qquad \textbf{(6.36)}$$

and a fourth-order Runge–Kutta method used, the values of y and z are obtained accurately only if the mesh spacing h is very small. Thus, for $h = .00055$, we find after 500 steps that $y(.275) = .75959$, which compares well with the exact value of .75957. With $h = .00275$ and 100 steps, we obtain $y(.275) = .75967$, also quite acceptable. However, with $h = .003$, only a slight increase, the value of y after 100 steps equals 2.2×10^9. For larger h the instability is worse. Since we had expected that even a value of h as large as .1 would have been acceptable, this result is disconcerting. What went wrong?

Since the Runge–Kutta method is algebraically complicated, let us apply the Euler method of Section 6.2 (for which a similar computational disaster occurs) to Equations (6.36). The computational process is described by

$$y_{j+1} - y_j = hz_j ,$$
$$z_{j+1} - z_j = -1001hz_j - 1000hy_j$$

or, in matrix form (Chapter 2), by

$$\begin{bmatrix} y_{j+1} \\ z_{j+1} \end{bmatrix} = \begin{bmatrix} 1 & h \\ -1000h & (1 - 1001h) \end{bmatrix} \cdot \begin{bmatrix} y_j \\ z_j \end{bmatrix} . \qquad \textbf{(6.37)}$$

In the Euler method we start with $y_0 = 1$, $z_0 = -1$ and apply Equation (6.37) for $j = 0, 1, 2, \ldots$ in sequence. Already we can see what would happen; unless the terms $1000h$ and $1001h$ occurring in Equation (6.37) are small, the square matrix will have large entries, and the successive multiplications by this matrix will lead to unreasonable results. This is only a qualitative argument, because there is always the possibility of self-cancelling, at least for appropriate initial conditions; consequently, it is worthwhile to look at Equation (6.37) more carefully. We now get a chance to use the eigenvalue discussion of Section 2.7. Denoting the square matrix in Equation (6.37) by A, construct an eigenvector matrix E as in Section 2.7 with the property that $AE = ED$, where D is a diagonal matrix whose diagonal entries are the two eigenvalues of A. Introduce new variables u_j and v_j by means of

$$\begin{bmatrix} y_j \\ z_j \end{bmatrix} = E \begin{bmatrix} u_j \\ v_j \end{bmatrix} . \tag{6.38}$$

(If desired, this equation may be solved for u_j, v_j in terms of y_j, z_j by multiplying both sides by E^{-1}.) Then Equation (6.37) becomes

$$E \begin{bmatrix} u_{j+1} \\ v_{j+1} \end{bmatrix} = AE \begin{bmatrix} u_j \\ v_j \end{bmatrix} = ED \begin{bmatrix} u_j \\ v_j \end{bmatrix} ,$$

so that

$$\begin{bmatrix} u_{j+1} \\ v_{j+1} \end{bmatrix} = D \begin{bmatrix} u_j \\ v_j \end{bmatrix} ,$$

Thus,

$$u_{j+1} = \lambda_1 u_j , \qquad v_{j+1} = \lambda_2 v_j ,$$

whose solution is

$$u_j = \lambda_1^j u_0 , \qquad v_j = \lambda_2^j v_0 , \tag{6.39}$$

where u_0 and v_0 correspond to y_0 and z_0 as given by Equation (6.38). Knowing u_j and v_j from Equation (6.39), Equation (6.38) now gives y_j and z_j; each will be a certain linear combination of the terms $\lambda_1^j u_0$ and $\lambda_2^j v_0$.

By direct calculation for the matrix A, we find

$$\lambda_1 = 1 - h , \qquad \lambda_2 = 1 - 1000h . \tag{6.40}$$

Because y_j and z_j possess terms involving the jth powers of λ_1 and λ_2, it is clear that y_j and z_j can become very large unless $|\lambda_1| \le 1$ and $|\lambda_2| \le 1$, and from Equations (6.40) this would require $h \le 1/500$ for this Euler method.

It could be argued that the coefficients of the power λ_2^j in y_j and z_j might be zero, perhaps because of the initial condition. However, even if the coefficient was exactly zero, the ubiquitous roundoff villain is always introducing minute amounts of each of the u_j, v_j components, and once this process starts, explosive growth would take over unless, indeed, $500h \le 1$.

The restriction on the step size is thus seen to be tied to the presence of the large eigenvalue. In terms of the original problem, whose general solution is a linear combination of e^{-x} and e^{-1000x}, it can equivalently be said that the restriction on h arises from the existence of two very different "scales" of x in the general solution. Even though the particular solution of interest was e^{-x} alone, the structure of the differential equation and so of the solution method still permits the possibility that the e^{-1000x} solution (with its much restrictive h requirement for stability) will be activated. Problems of this kind, which involve very different scales (time scales, if x is time) are termed *stiff*. Boundary layer problems [like that of Eq. (6.34)] are sometimes included in the "stiff" category because of the different rates of change of the solution inside and outside the boundary layer.

One way to reduce the mesh spacing difficulty for a stiff problem is to use an implicit method. We have previously remarked that implicit methods are less subject to stability constraints. In the present case a "backwards" Euler method (in which derivatives are evaluated at $j + 1$ rather than j) gives

$$y_{j+1} - y_j = hz_{j+1},$$

$$z_{j+1} - z_j = -1001hz_{j+1} - 1000hy_{j+1},$$

or

$$\begin{bmatrix} 1 & -h \\ 1000h & (1 + 1001h) \end{bmatrix} \begin{bmatrix} y_{j+1} \\ z_{j+1} \end{bmatrix} = \begin{bmatrix} y_j \\ z_j \end{bmatrix}.$$

Denoting the square matrix by B and introducing again an appropriate $E^{(1)}$ matrix such that $BE^{(1)}$ becomes diagonal, a straightforward calculation along the previous lines now leads to the result that the two constituents of y_j and z_j (corresponding to the previous u_j, v_j) will now decrease in magnitude as j grows. The process is stable, and roundoff is not a problem. For the problem of Equation (6.35), this backwards Euler method gives accurate results for $h = .01$, and even for $h = .1$ the error is only about 1%, over the interval $(0, 1)$.

If the initial conditions in the problem of Equation (6.35) are altered so that the exact solution now contains a combination of the terms e^{-x} and e^{-1000x}, then although the implicit method would continue to be stable for even fairly large h, it would be necessary to use a small value of h over that part of the interval where the e^{-1000x} term was significant, in order that the sharp rate of change of the solution corresponding to this term would be correctly modeled. Thus, it is generally desirable in stiff problems to use an adaptive mesh method, in which the local error is monitored and the mesh step adjusted accordingly. A simple implicit scheme [already met in Eq. (6.16)] that often works well is the *trapezoidal method*, defined for the equation $y' = f(x, y)$ by

$$y_{j+1} = y_j + \frac{h}{2} [f(x_j, y_j) + f(x_{j+1}, y_{j+1})] . \qquad (6.41)$$

To solve the nonlinear equation for y_{j+1} that is encountered at each step, a (perhaps modified) Newton method of the kind discussed in Chapter 1 is often recommended (cf. Prob. 6.21b).

6.7 EIGENVALUE PROBLEMS

In vibration type problems, or in the solution of a partial differential equation by the expansion of an unknown function in terms of simpler functions, one frequently encounters a kind of problem typified by the following simple example. Let $y(x)$ satisfy the differential equation

$$y'' + \lambda y = 0 \qquad (6.42)$$

in the interval $(0, 1)$, with $y(0) = 0$ and $y(1) = 0$, where λ is a constant determined by the condition that this problem is to have a solution other than the trivial one $y \equiv 0$.

The general solution of Equation (6.42) is given by $y = A \cos \sqrt{\lambda} x + B \sin \sqrt{\lambda} x$, where A and B are constants. If y is to vanish at $x = 0$, we must set $A = 0$. Thus, $y = B \sin \sqrt{\lambda} x$, and if $y = 0$ at $x = 1$, either B must vanish or $\sin \sqrt{\lambda} = 0$. Thus, a nontrivial solution can exist only if $\sin \sqrt{\lambda} = 0$, which corresponds to one of the choices $\lambda_n = n^2 \pi^2$ for λ, where n is any nonzero integer. Such a value for λ is termed an *eigenvalue*, and the corresponding solution function $y_n = B \sin n\pi x$ is termed an *eigenfunction*. Observe that B is arbitrary; any eigenfunction is determined only within an arbitrary multiplicative constant.

More complicated eigenvalue problems commonly have a similar structure, in that there is a sequence of possible eigenvalues, and as one moves further out in the sequence (i.e., as n increases), the corresponding eigenfunctions tend to make more oscillations within the interval of interest. The boundary conditions are usually somewhat more complicated than those in the example problem of Equation (6.42), although they are still homogeneous. We might, for example, require certain linear combinations of y and y' to vanish at the ends. A numerical approach might be as follows. Suppose, more generally, that

$$y'' + p(x) y' + [q(x) + \lambda r(x)] y = 0 \qquad (6.43)$$

for x in (a, b), where $p(x)$, $q(x)$, $r(x)$ are given functions of x. We are asked to find values of λ, say λ_n, for which nonidentically zero solutions to Equation (6.43) exist satisfying $y(a) = y(b) = 0$ (the method can be easily extended to more general homogeneous boundary conditions).

Since any solution will involve an arbitrary multiplicative constant, we can always set $y'(a) = 1$. For any choice of λ the solution y to Equation (6.43) can then be constructed numerically, and the value of y at $x = b$ recorded. This endpoint value of y is then, in principle, a function of λ, and we again have a problem of the kind discussed in Chapter 1, in which that value of λ for which $y(b)$ vanishes is sought.

Alternatively, Equation (6.43) can be replaced by a set of algebraic equations resulting from discretization over a set of mesh points, as in Section 6.5. A typical equation, with obvious notation, is

$$\frac{y_{j+1} - 2y_j + y_{j-1}}{h^2} + p_j \frac{y_{j+1} - y_{j-1}}{2h} + [q_j + \lambda r_j] y_j = 0$$

for $j = 1, 2, \ldots, n - 1$, with $y_0 = 0$, $y_n = 0$, if there are n subintervals. This represents a set of $n - 1$ linear homogeneous equations in the $n - 1$ unknowns, and for a nontrivial solution to exist the determinant of the coefficients must vanish. This results in a polynomial equation for λ that has $n - 1$ roots. One can expect that the first few values of λ are given reasonably well by this approach. To find the higher values of λ, one would have to use enough mesh points so that the corresponding highly oscillatory eigenfunctions can be well represented by the discretization.

6.8 BEST FIT METHODS

For some purposes an approximate analytical representation for the solution of a differential equation is desired. One approach would then be to choose a number of (independent) functions $\phi_1(x)$, $\phi_2(x)$, $\ldots$, $\phi_n(x)$ and to express the desired solution function $y(x)$ in terms of the $\phi_j(x)$ by

$$y(x) = c_1\phi_1(x) + c_2\phi_2(x) + \cdots + c_n\phi_n(x) . \qquad \textbf{(6.44)}$$

The constants c_j must be chosen so that this expression best fits (in some sense) the given differential equation and its initial or boundary conditions.

The phrase "best fits" can be interpreted in many ways, and its is useful to consider a specific example.

The velocity V of a moving body with a mass M (perhaps an automobile) changes because of the effects of propulsive force and drag. Let

$$M \frac{dV}{dt} = F(t) - AV - BV^2 ,$$

where t is time, $F(t)$ is the time-dependent applied force, and A, B are positive constants. Suppose that, after nondimensionalization, a particular situation is described by

$$\frac{dV}{dt} = 1 + t^2 - .2V - .3V^2 \qquad \textbf{(6.45)}$$

with $V(0) = 0$. The problem is to find a description of V in the interval $0 < t < 1$.

One possibility is to express V as a polynomial in t and to write

$$V = c_1 t + c_2 t^2 + c_3 t^3 , \tag{6.46}$$

say. The term in c_0 was omitted because of the initial condition $V(0) = 0$. If dV/dt is calculated, the discrepancy between the two sides of Equation (6.45) is·

$$E(t) = [c_1 + 2c_2 t + 3c_3 t^2]$$
$$- [1 + t^2 - .2(c_1 t + c_2 t^2 + c_3 t^3) - .3(c_1 t + c_2 t^2 + c_3 t^3)^2]. \tag{6.47}$$

There are now several approaches.

Least Squares Method

Choose the c_i so that $\int_0^1 E^2(t)\, dt$ is a minimum. As in the least squares discussion of Chapter 5, this leads to a set of three algebraic equations for the c_i.

Galerkin Method

Here one chooses a set of functions, say $\psi_i(t)$, and makes $E(t)$ orthogonal to each of these, in the sense (cf. Chap. 5)

$$\int_0^1 E(t)\psi_i(t)\, dt = 0 .$$

Since there are here three c_i to determine, it is appropriate to choose three ψ_i functions, and we might take $\psi_1 = 1$, $\psi_2 = 1 - t^2$, $\psi_3 = e^{-t}$. A weighting function could be inserted in the integral if a portion of the interval is of particular importance. If the $\psi_i(t)$ are the same as the "basis" functions used in Equation (6.46), that is, t, t^2, t^3, then the Galerkin method becomes identical to the least squares method.

Collocation Method

We can choose three points at which the differential equation is to be exactly satisfied by the form (6.46). If these three values of t are denoted by t_1, t_2, t_3, then, using Equation (6.47), we write $E(t_1) = 0$, $E(t_2) = 0$, $E(t_3) = 0$, and use these three conditions to determine the c_i values. For the reasons given in Chapter 5, it may be advantageous to choose, for the collocation points, the zeros of an appropriate Chebyshev polynomial.

In this example we chose simple polynomials for the ϕ_i functions of Equation (6.44). There are, however, many possibilities. The functions ϕ_i could, for example, be orthogonal polynomials, they could be trigonometric functions (e.g., part of a Fourier series), or special splines. Finlayson (1980) gives many examples in a chemical engineering context.

ANNOTATED BIBLIOGRAPHY

F. S. Acton, 1970, *Numerical Methods that Work*, Harper & Row, New York, 539p.

Chapters 5 and 6 give a treatment of numerical methods for differential equations that is readable, unconventional, instructive, and well supplied with drawings.

G. Dahlquist and Å. Björck, 1974, *Numerical Methods*, Prentice-Hall, Englewood Cliffs, New Jersey, 571p. (Translation of 1979 Swedish edition.)

The book offers an excellent treatment of topics in differential equations. Dahlquist has made important contributions to stability theory.

H. T. Davis, 1962, *Introduction to Nonlinear Differential and Integral Equations*, Dover, New York, 566p.

In solving nonlinear equations numerically, it can be very helpful to have some experience with the properties of such equations. This very readable book helps provide this kind of perspective.

B. Finlayson, 1980, *Nonlinear Analysis in Chemical Engineering*, McGraw-Hill, New York, 412p.

This text discusses a number of techniques for the solution of ordinary differential equations and illustrates them with practical examples.

C. W. Gear, 1971, *Numerical Initial Value Problems in Ordinary Differential Equations*, Prentice-Hall, Englewood Cliffs, New Jersey, 253p.

The treatment of stiff equations (Chap. 11), is particularly extensive.

L. W. Johnson and R. D. Riess, 1982, *Numerical Analysis*, 2nd ed., Addison-Wesley, Reading, Mass., 563p.

Chapter 7 discusses differential equations and gives a particularly thorough account of stability theory.

PROBLEMS

6.1 Show that a third-order Taylor series method (see Sec. 6.2) for $y' = f(x, y)$ has the form

$$y_{j+1} = y_j + hf + \tfrac{1}{2}h^2(f_x + f_y f)$$
$$+ \tfrac{1}{6}h^3(f_{xx} + 2f_{xy}f + f_{yy}f^2 + f_x f_y + f_y^2 f) ,$$

where f and its various partial derivatives (indicated by subscripts) are evaluated at x_j, y_j. If $f = x^2 - y^2$, obtain the coefficient of h^3 explicitly. Determine the tuncation error associated with this equation.

6.2 Let $y'' + xy' + y = 1$, where $y(0) = 1$, $y'(0) = 1$. Write this as a coupled pair of equations and use an appropriate computer subroutine (for the second-order Taylor process) to solve the problem in the interval (0, 1). Use Richardson extrapolation to improve the value of $y(1)$.

6.3 Find the truncation error of the implicit scheme

$$y_{j+1} - y_j = hf \left(\tfrac{1}{2}[x_i + x_{i+1}], \tfrac{1}{2}[y_j + y_{j+1}]\right)$$

for the problem $y' = f(x, y)$.

6.4 It is sometimes suggested that roundoff error can be largely minimized in, say, a Taylor series method for $y' = f(x, y)$ by computing f in single precision to save computer time, but by accumulating the y_j values in double precision. Experiment with this idea for the sample problem of Section 6.2, and evaluate the idea.

6.5 A third-order Runge–Kutta method for $y' = f(x, y)$ has the form

$$k_1 = f(x_j, y_j) ,$$
$$k_2 = f(x_j + \tfrac{1}{3}h, y_j + \tfrac{1}{3}hk_1) ,$$
$$k_3 = f(x_j + \tfrac{2}{3}h, y_j + \tfrac{2}{3}hk_2) ,$$
$$y_{j+1} = y_j + \frac{h}{4}(k_1 + 3k_3) .$$

Verify that these coefficients are appropriate.

6.6 Write a subroutine implementing the implicit Runge–Kutta method of Equations (6.25), using a nonlinear equation solver for k_2 that does not require differentiation of f, and that uses a fixed number of iterations. Experiment with the example problem $y' = x^2 - y^2$, $y(0) = 1$.

6.7 The subroutine RUKU of Section 6.3 can be made somewhat more efficient by using a subroutine, rather than the function FUN, to provide the array of derivative values. Rewrite the program to do this, and also permit the caller to include the name of the subroutine in the parameter list. Use the subroutine to experiment with $y'' + y = x$, $y(0) = 1$, $y'(0) = 0$ (which has the exact solution $y = x + \cos x - \sin x$) in the interval (0, 1), using both positive and negative values for h. At what power of h (if any) does the error at $x = 1$ tend to decrease as the number of steps increases? How well does Richardson extrapolation work in this case?

6.8 Apply Equation (6.27) to the special case $y' = -y$, $y(0) = 1$ (starting with $y_0 = 1$ and y_1 obtained from a Taylor expansion). Choose various step sizes and compute y_j. Verify the statements made in connection with Equation (6.27).

6.9 A popular predictor–corrector algorithm is *Milne's method*, defined by

(explicit) $\quad y_{j+1} = y_{j-3} + \dfrac{4h}{3} [2f(x_j, y_j) - f(x_{j-1}, y_{j-1}) + 2f(x_{j-2}, y_{j-2})]$,

(implicit) $\quad y_{j+1} = y_{j-1} + \dfrac{h}{3} [f(x_{j+1}, y_{j+1}) + 4f(x_j, y_j) + f(x_{j-1}, y_{j-1})]$.

Obtain the truncation errors. (The method is only weakly stable.)

6.10 Discuss, from the viewpoints of truncation error and stability, *Simpson's rule*, described by

$$y_{j+1} = y_{j-1} + \dfrac{h}{3} [f(x_{j+1}, y_{j+1}) + 4f(x_j, y_j) + f(x_{j-1}, y_{j-1})] .$$

6.11 A particular nonlinear oscillator is governed by the equation

$$\dfrac{d^2x}{dt^2} + \omega^2 x = \epsilon x^3 ,$$

where $x(t)$ represents the displacement, t is time, and ω and ϵ are constants. Let $x(0) = 1$, $x'(0) = 0$. Take $\omega = 1$ and solve this equation numerically for $\epsilon = .01$, $.1$, 1, 10, over a time interval that corresponds to some results of interest. What checks are applicable to your program?

6.12 A pair of equations governing the interaction of two populations (predator, prey) is given by

$$\dfrac{dP}{dt} = \alpha QP - \beta P , \qquad \dfrac{dQ}{dt} = \gamma - \delta Q - \tau P ,$$

where α, β, γ, δ, and τ are constants. Write a program that will solve this problem, given initial values of P and Q and the values of the parameters. Choose some reasonable values for the coefficients (note that the product QP arises from a product of P^2 and Q/P, the first term representing birth rate, which is proportional to the number of pairs) and discuss the nature of the solution.

6.13 The following problem arises in astronomy in connection with the gravitational field of a white dwarf: Find $y(x)$ satisfying

$$xy'' + 2y' + x(y^2 - c)^{3/2} = 0 ,$$

where $y(0) = 1$, $y'(0) = 0$. Solve it numerically in the range $0 < x < 5$ for $c = .1$. [*Hint*: The fact that the coefficient of y'' vanishes at $x = 0$ is a little troublesome. Divide through by x and use L'Hospital's rule on y'/x as $x \to 0$.]

6.14 Some "flip-flop" circuits are governed by Van der Pol's equation, which has the form

$$y'' - \epsilon(1 - y^2)y' + y = 0 ,$$

where ϵ is a constant and y is a function of time t. Take $y(0) = -3$, $y'(0) = 2$, and for each of the cases $\epsilon = .01, .1, 5$, obtain a solution curve for the interval $0 < t < 20$. Next, observe that the equation is *autonomous*, in that t does not occur explicitly; this raises the possibility of a "phase plane" approach. Define $w = y'$ and note that, since

$$y'' = \frac{d(y')}{dt} = \frac{d(y')}{dy} \frac{dy}{dt} ,$$

the equation can be rewritten as

$$\frac{dw}{dy} \cdot w - \epsilon(1 - y^2)w + y = 0 ,$$

which is the first-order equation relating w to y (or vice versa if dw/dy becomes large). Obtain a family of solutions for different initial conditions for this equation in the (y, w) plane for $\epsilon = 1$. Are there any interesting general features of these solution curves?

6.15 The following is an example of a pursuit problem. Let both pursuer and pursued move in the (x, y) plane; the pursued moves along the line $y = 1$ with unit velocity starting at time $t = 0$ at $x = 0$. Let the pursuer start from the origin and always move in the direction of the pursued; the pursuer moves with velocity 2 and starts at $t = 1$ (see Fig. 6.3). Verify that if $x(t)$, $y(t)$ are the coordinates of the pursuer, the equations of motion are

$$\dot{x}^2 + \dot{y}^2 = 4 , \qquad \frac{\dot{y}}{\dot{x}} = \frac{1 - y}{t - x} .$$

Solve numerically to find the time and place of interception.

6.16 Complete the solution of the problem of Equation (6.32), with $y(0) =$

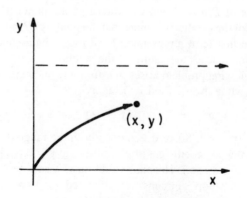

Figure 6.3. Pursuit problem.

1, $y(1) = 2$, using the shooting method. How many solutions can you find? Are there any solutions for the case $y(0) = 1$, $y(1) = 2$?

6.17 Complete the solution procedure associated with Equation (6.33). Try a similar method for Problem 6.16.

6.18 Sketch the exact solution of $\epsilon y'' - ky = -1$, $y(-1) = 0$, $y(1) = 0$ for $\epsilon = 10^{-6}$ and $k = 2.5$. [This problem relates to Eq. (6.34).]

6.19 Write an adaptive mesh program that utilizes Equation (6.41), and apply it to the problem of Equation (6.35), experimenting with various initial conditions.

6.20 Show that for the case $y' = \lambda y$, where λ is a constant, the Runge–Kutta method of Equations (6.25) becomes

$$y_{j+1} = y_j \left[1 + h\lambda + \frac{(h\lambda)^2}{2} + \frac{(h\lambda)^3}{6} + \frac{(h\lambda)^4}{24} \right].$$

The quantity in brackets coincides with the first few terms of the series expansion for $\exp(h\lambda)$; it is the amount by which each y_j is multiplied to obtain the succeeding y_{j+1}. If λ is negative, then y_j should decrease as j increases; show that this requires that h satisfy $|h\lambda| < 2.785$.

6.21 *a:* Consider using

$$y_{j+1} = y_j + \frac{h}{2} [3f_j - f_{j-1}]$$

from the Adams–Bashforth formulas to solve $y' = \lambda y$, where λ is a negative constant. Denote the two roots of

$$r^2 = r\left(1 + \frac{3h\lambda}{2}\right) - \frac{h\lambda}{2}$$

by r_1, r_2, and let r_1 be that particular root that (as in the convergence proof of Sec. 6.2) corresponds to the true solution. Explain why we must choose h such that $|r_2| \leqslant |r_1|$, and show that this condition (termed *relative stability*) leads to the constraint $h\lambda > -\frac{2}{3}$. *b:* The first of the Adams–Moulton formulas is sometimes called the *trapezoidal rule* because of the relationship to integration formula of the same name discussed in Chapter 7. Suppose that successive substitution is used in this formula to solve for y_{j+1} (cf. Section. 1.5). Show that the process will converge only if h is such that

$$\left| \frac{h}{2} \frac{\partial f}{\partial y} \right| < 1 ,$$

where $\partial f/\partial y$ is evaluated at (x_{j+1}, y_{j+1}). (We remark that for stiff problems, this kind of constraint implies that a Newton method is preferable.)

6.22 Let the velocity of light in a certain medium, relative to its velocity in a vacuum, be denoted by c. For a ray of light moving in the (x, y) plane, we have $c = c(x, y)$. Let s denote distance along a particular light ray; then the path of light ray will be defined by a specification of the x- and y-coordinates of any point on this path as functions of s. From optics the differential equations governing $x(s)$ and $y(s)$ are

$$\frac{d^2x}{ds^2} = -\frac{1}{c}\frac{\partial c}{\partial x} + \frac{1}{c}\frac{dx}{ds}\left[\frac{\partial c}{\partial x}\frac{dx}{ds} + \frac{\partial c}{\partial y}\frac{dy}{ds}\right],$$

$$\frac{d^2y}{ds^2} = -\frac{1}{c}\frac{\partial c}{\partial y} + \frac{1}{c}\frac{dy}{ds}\left[\frac{\partial c}{\partial x}\frac{dx}{ds} + \frac{\partial c}{\partial y}\frac{dy}{ds}\right].$$

Let $c = 1(1 + x^2)$, and let $x(0) = 0$, $y(0) = 0$, $x'(0) = 1\sqrt{2}$, $y'(0) = 1/\sqrt{2}$, for a particular light ray. Determine (numerically) the path of this ray for $0 < x < 1$. [*Note:* We require, of course, that $(dx/ds)^2 + (dy/ds)^2 = 1$. If this relation is satisfied initially, then the above differential equations guarantee that it will always be satisfied. Why?]

NUMERICAL DIFFERENTIATION AND INTEGRATION

7.1 NUMERICAL DIFFERENTIATION

There are many cases in which it is desirable to estimate the derivative of a function whose values are known only on a set of discrete points. For example, the electrostatic potential of a charge distribution may have been measured at a number of points; the electric field would then be obtained by differentiation. Similarly, the stresses in an elastic body are related to the derivatives of the displacement field. In numerical analysis this kind of problem is encountered in the replacement of an ordinary or partial differential equation by a finite difference equation, where derivatives at a mesh point must be expressed in terms of mesh point values of the unknown function. (One example of this was already considered in Section 6.5).

The simplest case is that in which $f(x)$ is known at x_1 and x_2 and in which $f'(x_0)$ is desired, where x_0 is some point near x_1 and x_2 (see Fig. 7.1a). From a Taylor expansion around the point x_0 (Appendix A), we have

$$f(x_1) = f(x_0) + (x_1 - x_0)f'(x_0) + \tfrac{1}{2}(x_1 - x_0)^2 f''(x_0) + \cdots ,$$

$$f(x_2) = f(x_0) + (x_2 - x_0)f'(x_0) + \tfrac{1}{2}(x_2 - x_0)^2 f''(x_0) + \cdots , \qquad \textbf{(7.1)}$$

providing $f(x)$ is sufficiently differentiable. (As usual, a prime denotes a derivative.) Subtracting the first equation of (7.1) from the second, we obtain

$$f'(x_0) = \frac{f(x_2) - f(x_1)}{x_2 - x_1} - \left(\frac{x_1 + x_2}{2} - x_0 \right) f''(x_0) + \cdots . \qquad \textbf{(7.2)}$$

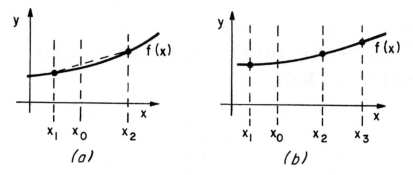

Figure 7.1. *a*: Estimate of $f'(x_0)$. *b*: Estimate of $f''(x_0)$.

Thus, $f'(x_0)$ can be approximated by $[f(x_2) - f(x_1)]/(x_2 - x_1)$, the dominant part of the error being given by

$$\left(\frac{x_1 + x_2}{2} - x_0\right) f''(x_0) . \tag{7.3}$$

The factor multiplying $f''(x_0)$ is equal to the distance between x_0 and the midpoint of the (x_1, x_2) interval. If $x_2 - x_1 = h$, this factor is some multiple of h, and we say that the error term (7.3) is of order h. We obtain the highest accuracy if x_0 is the midpoint of the interval (x_1, x_2), for then the expression (7.3) vanishes, and the error is of higher order. As a special case, x_0 may coincide with x_1 or x_2; the coefficient of $f''(x_0)$ in (7.3) is then $h/2$ or $-h/2$, respectively. We remark that a conventional notation for a term of order h^n is $O(h^n)$; this is the rate at which the error approaches zero, as h is decreased toward zero.

It is not reasonable to try to estimate $f''(x_0)$ from a knowledge of $f(x_1)$ and $f(x_2)$ alone, since the pair of equations (7.1) can be used to solve for only two unknowns, and for small h the terms in $f(x_0)$ and $f'(x_0)$ are dominant. To find an approximation for $f''(x_0)$, we need at least three points, so suppose (Fig. 7.1*b*) that $f(x_1)$, $f(x_2)$, $f(x_3)$ are known. Then expanding again about a point x_0 (which as a special case could coincide with x_1, x_2, or x_3) we have

$$f(x_1) = f(x_0) + (x_1 - x_0)f'(x_0) + \tfrac{1}{2}(x_1 - x_0)^2 f''(x_0)$$
$$+ \tfrac{1}{6}(x_1 - x_0)^3 f'''(x_0) + \cdots ,$$

$$f(x_2) = f(x_0) + (x_2 - x_0)f'(x_0) + \tfrac{1}{2}(x_2 - x_0)^2 f''(x_0)$$
$$+ \tfrac{1}{6}(x_2 - x_0)^3 f'''(x_0) + \cdots ,$$

$$f(x_3) = f(x_0) + (x_3 - x_0)f'(x_0) + \tfrac{1}{2}(x_3 - x_0)^2 f''(x_0)$$
$$+ \tfrac{1}{6}(x_3 - x_0)^3 f'''(x_0) + \cdots . \tag{7.4}$$

A pair of equations in which $f(x_0)$ does not appear can be obtained by subtracting the first equation of (7.4) from the second, and the second from the third. Manipulation of the resulting two equations gives

$$f'(x_0) = \frac{x_2 + x_3 - 2x_0}{(x_2 - x_1)(x_3 - x_1)} [f(x_2) - f(x_1)]$$

$$- \frac{x_1 + x_2 - 2x_0}{(x_3 - x_1)(x_3 - x_2)} [f(x_3) - f(x_2)]$$

$$+ \tfrac{1}{6} f'''(x_0)[(x_1 - x_0)(x_2 - x_0) + (x_2 - x_0)(x_3 - x_0)$$

$$+ (x_3 - x_0)(x_1 - x_0)] + \cdots , \tag{7.5}$$

and

$$f''(x_0) = 2 \frac{f(x_3) - f(x_2)}{(x_3 - x_2)(x_3 - x_1)} - 2 \frac{f(x_2) - f(x_1)}{(x_2 - x_1)(x_3 - x_1)}$$

$$- \left(\frac{x_1 + x_2 + x_3}{3} - x_0\right) f'''(x_0) + \cdots . \tag{7.6}$$

In each of Equations 7.5 and 7.6 the first two terms represent the practical approximation, and the third term represents the dominant part of the error as the spacing between points becomes small. Our main interest here is in Equation (7.6), and it is seen that if x_0 coincides with the mean of the three points (x_1, x_2, x_3), the error term will be $O(h^2)$ instead of $O(h)$. The error term for $f'(x_0)$ in Equation (7.5) is $O(h^2)$, which is an improvement over that corresponding to Equation (7.2). Again, it is not reasonable to try to determine $f'''(x_0)$ by using only $f(x_1)$, $f(x_2)$, and $f(x_3)$.

If $x_3 - x_2 = x_2 - x_1 = h$, and if x_0 coincides with one of the points x_1, x_2, x_3, then as special cases of Equations 7.5 and 7.6 we obtain (denoting $f(x_2)$ by f_2, $f'(x_2)$ by f_2', etc.)

$$f_1' = \frac{4f_2 - 3f_1 - f_3}{2h} + O(h^2) ,$$

$$f_2' = \frac{f_3 - f_1}{2h} + O(h^2) ,$$

$$f_3' = \frac{f_1 + 3f_3 - 4f_2}{2h} + O(h^2) , \tag{7.7}$$

and

$$f_1'' = \frac{f_1 - 2f_2 + f_3}{h^2} + O(h) ,$$

$$f_2'' = \frac{f_1 - 2f_2 + f_3}{h^2} + O(h^2) ,$$

$$f_3'' = \frac{f_1 - 2f_2 + f_3}{h^2} + O(h) . \tag{7.8}$$

Note the $O(h^2)$ term in the second equation of (7.8).

Higher derivatives can be obtained similarly by using more points. It usually does not pay to try to determine a low-order derivative by using many points since the above process really involves polynomial interpolation, and high-order polynomials are notoriously prone to oscillations. Also, it is reasonable to expect that only information in the immediate neighborhood of x_0 should have much effect on, say f_0'.

The above formulas are also applicable to differentiable functions of a complex variable, although alternatives based on integral identities for analytic functions are also useful.

Approximations for partial derivations may be obtained similarly, using now the Taylor expansion formula for functions of several variables. Considering, for example, the case of a function f of two variables x, y, where f is known at equally spaced mesh points (see Fig. 7.2), we obtain (where $\partial f/\partial x|_0$ means $\partial f/\partial x$ evaluated at x_0, y_0, etc.)

$$\left.\frac{\partial f}{\partial x}\right|_0 = \frac{f_3 - f_1}{2h} + O(h^2) \ ,$$

$$\left.\frac{\partial f}{\partial y}\right|_0 = \frac{f_2 - f_4}{2h} + O(h^2) \ ,$$

$$\left.\frac{\partial^2 f}{\partial x^2}\right|_0 = \frac{f_1 - 2f_0 + f_3}{h^2} + O(h^2) \ ,$$

$$\left.\frac{\partial^2 f}{\partial y^2}\right|_0 = \frac{f_2 - 2f_0 + f_4}{h^2} + O(h^2) \ ,$$

$$\left.\left(\frac{\partial^2 f}{\partial x^2} + \frac{\partial^2 f}{\partial y^2}\right)\right|_0 = \frac{f_1 + f_2 + f_3 + f_4 - 4f_0}{h^2} + O(h^2) \ ,$$

$$\left.\frac{\partial^2 f}{\partial x\,\partial y}\right|_0 = \frac{f_6 - f_5 - f_8 + f_7}{4h^2} + O(h^2) \ , \tag{7.9}$$

and so on.

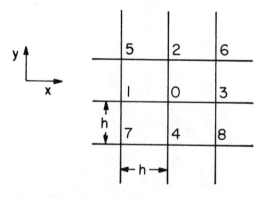

Figure 7.2. Mesh points for partial derivatives.

7.2 NUMERICAL INTEGRATION

The solutions of many practical problems are conveniently formulated in terms of integrals. One example arises in heat diffusion. If the time-dependent temperature field corresponding to a concentrated heat impulse at time zero is known, the temperature field resulting from any initial temperature distribution may be obtained by a superposition process and expressed as an integral. As in Fourier series, integrals also arise in the coefficient terms for spectral decomposition. The related method of integral transforms leads to solutions in the form of integrals. Also, some frequently encountered functions are most easily defined and evaluated by means of integrals (one example is the error function of Chap. 4). Since many of the integrals encountered in practice cannot be readily expressed in terms of simpler functions, it is worthwhile to develop efficient methods for numerical integration. Consider first

$$I = \int_a^b f(x) \, dx \, , \qquad (7.10)$$

where $f(x)$ is a continuous function in the interval (a, b). (If f has one or more discontinuities, we assume the integral can be written as a sum of integrals of the form (7.10), each over a subinterval in which f is continuous.) One way to evaluate I is by defining

$$y(\xi) = \int_a^\xi f(x) \, dx$$

so that y satisfies the differential equation

$$y'(\xi) = f(\xi) \qquad (7.11)$$

with $y(a) = 0$. The methods of Chapter 6 could be used to solve Equation (7.11), and then I would be obtained from $I = y(b)$. However, differential equation methods are designed to deal with the case in which f depends on both ξ and y, and Equation (7.11) is simpler than this, so that more effective procedures should be available.

Another approach is to choose a number of uniformly spaced mesh points in the interval, as shown in Figure 7.3. Let the values of $f(x)$ at the mesh points x_0, $x_1, \ldots, x_n$ be denoted by $f_0, f_1, \ldots, f_n$, respectively; the mesh spacing is h. Then we could interpolate f by a polynomial (or a spline) that agrees with f at the given mesh points and that can be readily integrated to give an approximation to $\int f$. On the other hand, I is the sum of integrals over the individual subintervals, and it would seem simpler to "fit" $f(x)$ locally on each subinterval rather than try to find a single representation valid over the entire interval.

This idea leads to the methods most commonly used. Consider any one subinterval, say between x_j and x_{j+1}; f will have values f_j and f_{j+1} at the two ends. A simple approximation to f over the subinterval is obtained by drawing a straight line between these endpoint values, as in Figure 7.4; the area of the resulting trapezoid is $\frac{1}{2} h(f_j + f_{j+1})$. Adding up similar results for all the subintervals, we obtain the *trapezoidal rule*

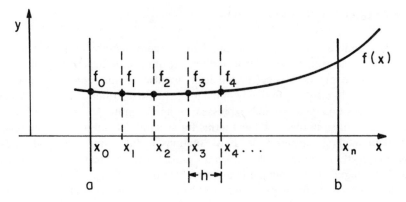

Figure 7.3. Interpolation points for evaluation of $\int_a^b f(x)\, dx$.

$$\int_a^b f(x)\, dx \cong h[\tfrac{1}{2} f_0 + f_1 + f_2 + \cdots + f_{n-1} + \tfrac{1}{2} f_n]. \tag{7.12}$$

It is useful to have an estimate for the error resulting from the use of Equation (7.12). This can be obtained if $f(x)$ is twice continuously differentiable. From Problem 5.2, $f(x)$ in the interval (x_j, x_{j+1}) can be written as

$$f(x) = f_j + (x - x_j)\frac{f_{j+1} - f_j}{h} + (x - x_j)(x - x_{j+1})\frac{f''(\xi)}{2!},$$

where ξ is some point in (x_j, x_{j+1}); ξ will depend on x. Integration of this equation gives

$$\int_{x_j}^{x_{j+1}} f(x)\, dx = \frac{1}{2} h(f_j + f_{j+1}) + \int_{x_j}^{x_{j+1}} (x - x_j)(x - x_{j+1})\frac{f''(\xi)}{2}\, dx.$$

The product $(x - x_j)(x - x_{j+1})$ is negative for all x in (x_j, x_{j+1}), so the righthand integral must be bounded by the two results obtained by replacing $f''(\xi)$ by the

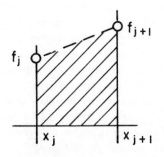

Figure 7.4. Trapezoid rule.

maximum and minimum values attained by f'' in the subinterval. This means that the value of this integral must equal some average value of $f''(\xi)$ times

$$\frac{1}{2} \int_{x_j}^{x_{j+1}} (x - x_j)(x - x_{j+1}) \, dx = -\frac{1}{12} h^3 .$$

Since f'' attains all values between its maximum and minimum values, there must be some point, say ζ, in the subinterval where the requisite average value is actually attained, so that

$$\int_{x_j}^{x_{j+1}} f(x) \, dx = \frac{1}{2} h(f_j + f_{j+1}) - \frac{1}{12} h^3 f''(\zeta) .$$

Now add up these results for all the subintervals. The error term for each subinterval involves a different value of f'', as attained at some point in each subinterval. Overall, we therefore have as the error term $-\frac{1}{12} h^3 (n f''_{av})$, where f''_{av} denotes the average of the $f''(\zeta)$ quantities. Again, f''_{av} is actually attained at some point ω in (a, b), so that

$$\int_b^a f(x) \, dx = h \left[\frac{1}{2} f_0 + f_1 + f_2 + \cdots + f_{n-1} + \frac{1}{2} f_n \right] - \frac{(b - a)h^2}{12} f''(\omega) ,$$

$$(7.13)$$

where ω is some point in (a, b), and where we have used $nh = b - a$. As an example, the value of

$$J = \int_0^\pi \sqrt{2 + \sin x} \, dx$$

using 10 subintervals would be given by the trapezoid rule with an error of less than .02. With $n = 100$, the error would be less than .0002. (The trapezoidal rule is pretty good.)

We can expect an improvement if $f(x)$ is fitted locally with a parabola rather than a straight line. This involves three interpolation points, so we use a pair of subintervals and the three points x_{j-1}, x_j, x_{j+1}. (This will require n to be even.) An analogous calculation now gives.

$$\int_a^b f(x) \, dx = \frac{h}{3} [f_0 + 4f_1 + 2f_2 + 4f_3 + 2f_4 + \cdots + 2f_{n-2} + 4f_{n-1} + f_n]$$

$$- \frac{1}{180} h^4 (b - a) f^{(4)}(\omega) ,$$

$$(7.14)$$

where again ω is some point in (a, b). Apart from the first and last coefficients, the coefficients in this *Simpson's rule* formula alternate between 4 and 2. Note that the error is $O(h^4)$ rather than $O(h^3)$ in Equation (7.14), as might be expected by comparison with Equation (7.13). The reason is that the interpolation term that would normally represent an error is odd with respect to the center of the double subinterval and so gives no contribution to the integral; see Problem 7.5.

The trapezoidal and Simpson's rules are special cases of the *Newton–Cotes formulas of closed type* (so called because each set of basic subintervals includes the endpoints). With obvious notation, the first few of these formulas (for the basic subinterval set) are

$$\int_{x_0}^{x_1} f(x)\,dx = \frac{h}{2}(f_0 + f_1) - \frac{h^3}{12}f''(\xi)\,,$$

$$\int_{x_0}^{x_2} f(x)\,dx = \frac{h}{3}(f_0 + 4f_1 + f_2) - \frac{h^5}{90}f^{(4)}(\xi)\,,$$

$$\int_{x_0}^{x_3} f(x)\,dx = \frac{3h}{8}(f_0 + 3f_1 + 3f_2 + f_3) - \frac{3h^5}{80}f^{(4)}(\xi)\,,$$

$$\int_{x_0}^{x_4} f(x)\,dx = \frac{2h}{45}(7f_0 + 32f_1 + 12f_2 + 32f_3 + 7f_4) - \frac{8h^7}{945}f^{(6)}(\xi)\,. \quad (7.15)$$

Others are given in Abramowitz and Stegun (1964). In these equations, and also in the next set, ξ is some point in the interval of integration.

The Newton–Cotes formulas of *open type* are similar, except that endpoint values are not used in the basic set of subintervals. The first few are

$$\int_{x_0}^{x_2} f(x)\,dx = 2hf_1 + \frac{h^3}{3}f''(\xi)\,,$$

$$\int_{x_0}^{x_3} f(x)\,dx = \frac{3h}{2}(f_1 + f_2) + \frac{h^3}{4}f''(\xi)\,,$$

$$\int_{x_0}^{x_4} f(x)\,dx = \frac{4h}{3}(2f_1 - f_2 + 2f_3) + \frac{14h^5}{45}f^{(4)}(\xi)\,,$$

$$\int_{x_0}^{x_5} f(x)\,dx = \frac{5h}{24}(11f_1 + f_2 + f_3 + 11f_4) + \frac{95h^5}{144}f^{(4)}(\xi)\,. \quad (7.16)$$

Not surprisingly, the open formulas do not give results as good as the closed ones. As partial compensation, they do not require f to be evaluated at as many points; also, they have an advantage if f is not defined at one or both endpoints of the parent interval. Some of the coefficients in the higher-order formulas may be negative (see Abramowitz and Stegun, 1964).

Richardson extrapolation may be used with any of these formulas. For the trapezoidal rule of Equation (7.13), write

$$I = T_1 + ch^2 ,$$

where I is the true value of the integral, T_1 the result obtained by using the trapezoidal rule with a mesh spacing of h, and c the coefficient in the error term of Equation (7.13). Repeating with a mesh spacing of $h/2$, and assuming that the same value of c is appropriate, we get

$$I = T_2 + c\left(\frac{h}{2}\right)^2 ,$$

where T_2 is the new trapezoidal result. The error term may not now be eliminated from these two equations to give

$$I \cong \tfrac{1}{3}(4T_2 - T_1) .$$

Actually, it can be shown (see Prob. 7.8 for the idea) that the error in this extrapolated value for I is now of order h^4, so we could do everything over again, starting with spacings of $h/2$ and $h/4$ and combining the results to get a new extrapolate. The two extrapolates so obtained differ from the true integral by a term kh^4, and Richardson extrapolation may now be applied to these two extrapolates. This kind of interated extrapolation, which may be repeated indefinitely, is called *Romberg integration*. It is sometimes used in high-precision calculations.

Any of the Newton–Cotes (or other) integration formulas may be used as the basis for an adaptive mesh computer program in which the error is monitored as the integration process proceeds (e.g., using a different formula or halved mesh), and the local mesh spacing adjusted appropriately. One input to such a subroutine would be the permissible overall error.

If a periodic function is to be integrated over its period, then the trapezoidal rule (which now involves equal coefficients, since f_0, and f_n coincide) is as good as anything. For if there were a better formula, the associated pattern (i.e., the weights) could be shifted any desired number of mesh points without altering its optimal character, and averaging over all such shifts gives the trapezoidal rule.

7.3 GAUSSIAN QUADRATURE

In 1814 Gauss suggested a method of integration that still delights anyone seeing it for the first time. The Newton–Cotes formulas use equal mesh intervals; Gauss asked, in effect, "Suppose we release the mesh points from this restriction. Where should they be placed for optimal accuracy?"

Consider the interval $(-h, h)$ over which a function $f(x)$ is defined, and suppose we use just two evaluation points, x_1 and x_2. Assume that $f(x)$ has a Taylor expansion about the origin, so that

$$f(x) = f(0) + xf'(0) + \tfrac{1}{2} x^2 f''(0) + \cdots .$$

Then the exact value of $I = \int_{-h}^{h} f(x)\, dx$ is given by

$$I = 2hf(0) + \frac{2}{3!} h^3 f''(0) + \frac{2}{5!} h^5 f^{(4)}(0) + \cdots .$$

We want to approximate I by an expression of the form

$$J = w_1 f(x_1) + w_2 f(x_2) ,$$

where x_1, x_2, and the coefficients w_1, w_2 are chosen so as to minimize the discrepancy between I and J. Now

$$J = w_1[f(0) + x_1 f'(0) + \tfrac{1}{2} x_1^2 f''(0) + \cdots]$$
$$+ w_2[f(0) + x_2 f'(0) + \tfrac{1}{2} x_2^2 f''(0) + \cdots] ,$$

so that, equating the coefficients of the various derivates of f in the two expressions I and J, we obtain

$$w_1 + w_2 = 2h , \qquad w_1 x_1 + w_2 x_2 = 0 ,$$

$$\frac{1}{2}[w_1 x_1^2 + w_2 x_2^2] = \frac{2}{3!} h^3 , \qquad \frac{1}{3!}[w_1 x_1^3 + w_2 x_2^3] = 0 .$$

There is no need to proceed further since these four equations determine x_1, x_2, w_1, and w_2. Their solution is $x_1 = -h/\sqrt{3}$, $x_2 = h/\sqrt{3}$, $w_1 = h$, $w_2 = h$. The discrepancy between I and J is a term involving $f^{(4)}(0)$, and we find easily, that

$$\int_{-h}^{h} f(x)\, dx = h\left[f\left(\frac{-h}{\sqrt{3}}\right) + f\left(\frac{h}{\sqrt{3}}\right) \right] + \frac{1}{135} h^5 f^{(4)}(0) + \cdots . \tag{7.17}$$

Thus, fourth-order accuracy is achieved with just two evaluation points!

To generalize this result, it is convenient to transform the variable so that the base interval becomes $(-1, 1)$. Given n, the problem is now to find $x_1, \ldots, x_n$ and the associated coefficients $w_1, w_2, \ldots, w_n$, so that the formula

$$\int_{-1}^{1} f(x)\, dx \cong w_1 f(x_1) + w_2 f(x_2) + \cdots + w_n f(x_n) \tag{7.18}$$

is exact when $f(x)$ is a polynomial of as high a degree as possible (for comparison with the preceding calculation, note that the first k terms of a Taylor expansion, for any choice of k, form a polynomial).

Recall from Section 5.5 that the Legendre polynomials, $P_n(x)$, are orthogonal over the interval $(-1, 1)$, with weight function unity (some properties of the $P_n(x)$ are listed in Problem 7.10). Denote the zeros of $P_n(x)$ by $x_1, x_2, \ldots, x_n$; from Problem 7.10 these are all real and distinct and lie in $(-1, 1)$. It will now be shown that if these particular (x_i) values are chosen in the integration formula (7.18), then Equation (7.18) is exact if $f(x)$ is a polynomial of order less than or equal to $2n - 1$. To prove this, observe first that the w_i are determined once and for all by the choice of the (x_i) points. Each w_i must in fact equal the integral over $(-1, 1)$ of the Lagrange polynomial (Sec. 5.1) associated with the point x_i if

Equation (7.18) is to be exact when $f(x)$ is a polynomial of order $n - 1$ or less. Now let $f(x)$ be a polynomial $\psi_m(x)$ of order $m > n - 1$, and denote by L_{n-1} that polynomial of order $n - 1$ that interpolates $\psi_m(x)$ at the points (x_i). Then $\psi_m - L_{n-1}$ is a polynomial of order m that vanishes at each point x_i and so must have the form $(x - x_1)(x - x_2) \cdots (x - x_n) r(x)$, where $r(x)$ is a polynomial of order $m - n$. Thus,

$$\int_{-1}^{1} \psi_m(x)\, dx = \int_{-1}^{1} L_{n-1}(x)\, dx$$

$$+ \int_{-1}^{1} (x - x_1)(x - x_2) \cdots (x - x_n) r(x)\, dx\,. \quad \textbf{(7.19)}$$

The first term on the righthand side of Equation (7.19) equals $w_1 \psi_m(x_1) + \cdots + w_n \psi_m(x_n)$. In the second term the factor $(x - x_1)(x - x_2) \cdots (x - x_n)$ equals $P_n(x)$, apart from a multiplicative constant, because of the choice of the x_i. Since $P_n(x)$ is orthogonal to any polynomial of order less than n, this second term vanishes if the order $m - n$ of $r(x)$ is less than n—that is, if $m \leq 2n - 1$—and the proof is now complete.

Now let $f(x)$ be any $2n$-times continuously differentiable function. It is useful to have an expression for the error term E in

$$\int_{-1}^{1} f(x)\, dx = w_1 f(x_1) + w_2 f(x_2) + \cdots + w_n f(x_n) + E\,. \quad \textbf{(7.20)}$$

We already know (because of the choice of x_i points) that the error term E vanishes if $f(x)$ is any polynomial of order $2n - 1$ or less. For the given function $f(x)$, choose as interpolation points the above x_i points and also a set of n points $x_i + \epsilon$, where $|\epsilon|$ is very small, and let $L_{2n-1}(x)$ be that polynomial that interpolates $f(x)$ at these $2n$ points. Then from Problem 5.2,

$$f(x) = L_{2n-1}(x)$$

$$+ (x - x_1) \cdots (x - x_n)(x - [x_1 + \epsilon]) \cdots (x - [x_n + \epsilon]) \frac{f^{(2n)}(\xi)}{(2n)!}\,,$$

where ξ is some point in $(-1, 1)$. Equation (7.20) is exact for the L_{2n-1} part of $f(x)$, so, allowing $\epsilon \to 0$, the value of E is given by

$$E = \int_{-1}^{1} (x - x_1)^2 (x - x_2)^2 \cdots (x - x_n)^2 \frac{f^{(2n)}(\xi)}{(2n)!}\, dx$$

(where ξ is some function of x, with $-1 < \xi < 1$). The factor multiplying the derivative term in the integrand is positive, so by the usual argument we deduce

$$E = \frac{f^{(2n)}(\zeta)}{(2n)!} \int_{-1}^{1} (x - x_1)^2 \cdots (x - x_n)^2\, dx\,,$$

where ζ is some point in $(-1, 1)$. The integrand is now proportional to P_n^2, and using the facts that the coefficient of x^n in P_n is

$$\frac{(2n - 1)!}{2^{n-1} n!(n - 1)!}$$

and that

$$\int_{-1}^{1} P_n^2 \, dx = \frac{2}{2n + 1},$$

we find

$$E = \frac{2^{2n+1}(n!)^4}{[(2n)!]^3(2n + 1)} f^{(2n)}(\zeta) \tag{7.21}$$

for the error term in Equation (7.20), where ζ is some point in $(-1, 1)$.

Abramowitz and Stegun (1964) tabulate the locations of the zeros for $P_n(x)$, and the associated w_i values, for selected values of n up to $n = 96$. Some of these are

$\pm x_i$	w_i
	$n = 2$
.57735 02691 89626	1.00000 00000 00000
	$n = 3$
.00000 00000 00000	.88888 88888 88889
.77459 66692 41483	.55555 55555 55556
	$n = 4$
.33998 10435 84856	.65214 51548 62546
.86113 63115 94053	.34785 48451 37454
	$n = 8$
.18343 46424 95650	.36268 37833 78362
.52553 24099 16329	.31370 66458 77887
.79666 64774 13627	.22238 10344 53374
.96028 98564 97536	.10122 85362 90376

If, instead of $(-1, 1)$, the interval is $(-h, h)$ (as would be the case if an original interval (a, b) is broken into subintervals of length $2h$), then the $\pm x_i$ values and the w_i values in the above table should be multiplied by h. The error term in Equation (7.21) is multiplied by h^{2n+1}, and ζ now becomes a point in $(-h, h)$. Thus, for $n = 4$,

$$E = (2.879) \, (10^{-7}) h^9 f^{(8)}(\zeta) \, .$$

The Gaussian quadrature idea is also applicable to the case of an integral involving a weight function $w(x)$. The reader can easily extend the argument given above to show that

$$\int_a^b w(x) f(x)\, dx = w_1 f(x_1) + \cdots + w_n f(x_n)$$

is exact, when $f(x)$ is a polynomial of order $\leq 2n - 1$, provided that the (x_i) are the zeros of the nth orthogonal polynomial (with respect to the weight function $w(x)$; see Sec. 5.5), and provided, of course, that the constants w_i are appropriately chosen. For the special case

$$\int_{-1}^1 \frac{1}{\sqrt{1 - x^2}}\, f(x)\, dx \ ,$$

the orthogonal polynomials are again the Chebyshev polynomials of Section 5.5.

Tabulated values of the x_i and w_i are given by Abramowitz and Stegun (1964) for a number of $w(x)$ and (a, b) values (including some infinite interval situations).

7.4 SOME INTEGRATION DIFFICULTIES

The methods of Sections 7.2 and 7.3 work best if the integrand (apart perhaps from some weighting factor) is well behaved—that is, bounded, continuous, and preferably possessing continuous derivatives of some order. Many integrals encountered in practice are not this nice. It may, of course, be possible to modify the integral, prior to the use of a numerical method, by writing the overall integral as the sum of a number of simpler integrals, or by a suitable change in variable (which, for example, changes an infinite interval into a finite one, or which removes a factor like $1/\sqrt{x}$ in the integrand by the transformation $x = t^2$, so that $dx/\sqrt{x}$ becomes just $2\, dt$). However, after reasonable manipulations of this kind have been tried, difficulties may well remain.

It often turns out that the range of integration is infinite. Consider, for example,

$$I = \int_0^\infty \frac{dx}{1 + x^4 + \sin x} \ .$$

We could write

$$\int_0^\infty = \int_0^A + \int_A^\infty$$

and choose A large enough so that the second term on the right—the "tail" of the integral—is small enough to be neglected. Since $|\sin| \leq 1$, we know

$$J = \int_A^\infty \frac{dx}{1 + x^4 + \sin x} < \int_A^\infty \frac{dx}{x^4} = \frac{1}{3A^3} \ ,$$

and this provides a criterion for the choice of A. If it turns out that A has to be uncomfortably large, the integral J could be integrated by parts to give

$$J = \left[\frac{x}{1 + x^4 + \sin x} \right]_A^\infty + \int_A^\infty \frac{x(4x^3 + \cos x)}{(1 + x^4 + \sin x)^2} \, dx$$

$$= \frac{-A}{1 + A^4 + \sin A} + \int_A^\infty \frac{4[x^4 + 1 + \sin x] + [x \cos x - 4 - 4 \sin x]}{(1 + x^4 + \sin x)^2} \, dx$$

$$= \frac{-A}{1 + A^4 + \sin A} + 4J + \int_A^\infty \frac{x \cos x - 4 - 4 \sin x}{(1 + x^4 + \sin x)^2} \, dx \ .$$

Thus,

$$3J = \frac{A}{1 + A^4 + \sin A} + \int_A^\infty \frac{4 + 4 \sin x - x \cos x}{(1 + x^4 + \sin x)^2} \ .$$

A bound on the integral is given by

$$\left| \int_0^A \frac{4 + 4 \sin x - x \cos x}{(1 + x^4 + \sin x)^2} \, dx \right| < \int_A^\infty \frac{4 + 4 + x}{x^8} \, dx$$

$$< \frac{8}{7A^7} + \frac{1}{6A^6} \ ,$$

so that, in terms of the original integral I, we now have

$$I = \int_A^\infty \frac{dx}{1 + x^4 + \sin x} + \frac{A}{3[1 + A^4 + \sin A]}$$

within an error of no more than $8/21A^7 + 1/18A^6$.

For a different kind of example, let $f(x)$ be well behaved and consider

$$I = \int_\epsilon^1 \frac{f(x)}{e^x - 1 - x} \ ,$$

where $0 < \epsilon \ll 1$. The denominator looks like $x^2/2$ for small x, so the integrand has a steep slope near the lower limit, ϵ, and is certainly not well fitted by a polynomial. Consequently, numerical integration formulas would require the use of many subintervals for accuracy. However, we can "subtract off the singularity," by writing

$$I = \int_\epsilon^1 \frac{f(x) - [f(0) + xf\,'(0)]}{e^x - 1 - x} \, dx + \int_\epsilon^1 \frac{f(0) + xf\,'(0)}{e^x - 1 - x} \, dx \ .$$

The first integrand is now bounded and presents no numerical difficulty. Rewrite the second as

$$\int_{\epsilon}^{1} [f(0) + xf'(0)] \left[\frac{1}{e^x - 1 - x} - \frac{2}{x^2} + \frac{2}{3x} \right] dx$$

$$+ \int_{\epsilon}^{1} \left(\frac{2}{x^2} - \frac{2}{3x} \right) [f(0) + xf'(0)] \, dx \, .$$

Again, the first integrand is bounded, and the second may be explicitly integrated to give

$$2f(0) \cdot \left(\frac{1}{\epsilon} - 1 + \frac{1}{3} \ln \epsilon \right) - 2f'(0) \cdot \left(\frac{1}{3} + \ln \epsilon - \frac{1}{3} \epsilon \right) .$$

Thus, the only numerical integrations required involve bounded smooth integrands.

The next example introduces the useful concept of an asymptotic expansion. Suppose one wants to compute the value of the complementary error function erfc(x) (cf. Chap. 4) to high accuracy. The definition is

$$\text{erfc}(x) = \frac{2}{\sqrt{\pi}} \int_{x}^{\infty} e^{-t^2} dt \, ,$$

and again there is the difficulty of estimating the effect of the "tail." Write $t = x + \tau$ to give

$$\text{erfc}(x) = \frac{2}{\sqrt{\pi}} e^{-x^2} \int_{0}^{\infty} e^{-2x\tau - \tau^2} d\tau \, .$$

For large x the factor $e^{-2x\tau}$ decays with exponential rapidity, so the major contribution to the integral must arise from the behavior of the integrand for small values of τ. Thus, the factor $e^{-\tau^2}$ can be approximated by its behavior near the origin, so

$$\text{erfc}(x) \sim \frac{2}{\sqrt{\pi}} e^{-x^2} \int_{0}^{\infty} e^{-2x\tau} \left[1 - \tau^2 + \frac{\tau^4}{2!} - \cdots \right] d\tau$$

$$\sim \frac{2}{\sqrt{\pi}} e^{-x^2} \left[\frac{1}{2x} - \frac{2!}{(2x)^3} + \frac{4!}{2!(2x)^5} - \cdots \right] ,$$

where the symbol $\sim$ means "is asymptotic to," or "is well represented by for large x." This looks like an infinite series, but it is not. For any chosen value of x, it would (quite clearly) ultimately diverge. However, it does have the useful property that if one stops after any fixed number of terms (say 2 or 3, for example), then as x becomes larger and larger, the sum of these terms represents the integral more and more accurately. (Moreover, the error in such an *asymptotic expansion* is usually of the order of the first omitted term). Using this result now to evaluate, say, erfc(2.3), we would write

$$\text{erfc}(2.3) = \frac{2}{\sqrt{\pi}} \int_{2.3}^{5} e^{-t^2} \, dt + \frac{2}{\sqrt{\pi}} e^{-25} \left[\frac{1}{10} - \frac{1}{500} - \cdots \right] ,$$

and the correction term to the numerically obtained integral is of order 10^{-12}.

7.5 MORE DIMENSIONS

It will be sufficient to consider the case of integration over a two-dimensional region, since higher-dimensional cases are analogous.

From calculus the integral of a function $f(x, y)$ over a region A of the plane can be calculated by first integrating with respect to x and then with respect to y, or vice versa; this fact provides one method for the evaluation of a two-dimensional integral. The process is particularly straightforward if A is a rectangle, defined say by $a \le x \le b, c \le y \le d$. Arbitrarily choosing the y-direction to first integrate along, we have

$$I = \int_{A} f(x, y) \, dA = \int_{a}^{b} dx \left(\int_{c}^{d} f(x, y) \, dy \right) . \tag{7.22}$$

In the integral $\int_{c}^{d} f(x, y) \, dy$, x plays the role of a parameter. This integral is a function of x and will be denoted by $Q(x)$. Then

$$I = \int_{a}^{b} Q(x) \, dx . \tag{7.23}$$

If $Q(x)$ were known, any of the standard algorithms could be applied to the numerical determination of I. For example, Simpson's rule gives

$$I \cong \frac{h}{3} \left[Q(x_0) + 4Q(x_1) + 2Q(x_2) + \cdots + Q(x_m) \right] . \tag{7.24}$$

Here $x_j = a + jh$ (see Fig. 7.5), where $b - a = mh$, m being an even integer. To find $Q(x_j)$, Simpson's rule can be used again, now with a mesh spacing k in the y-direction, so that $d - c = nk$ (with n even). Then

$$Q_j = \frac{k}{3} \left[f(x_j, y_0) + 4f(x_j, y_1) + 2f(x_j, y_2) + \cdots + f(x_j, y_n) \right] . \tag{7.25}$$

Combining Equations (7.24) and (7.25),

$$I \cong \frac{kh}{9} \left[f(x_0, y_0) + 4f(x_1, y_0) + 2f(x_2, y_0) + \cdots \right.$$

$$\left. + 4f(x_0, y_1) + 16f(x_1, y_1) + \cdots \right] . \tag{7.26}$$

The coefficient pattern for a corner of the rectangular region is shown in Figure 7.5.

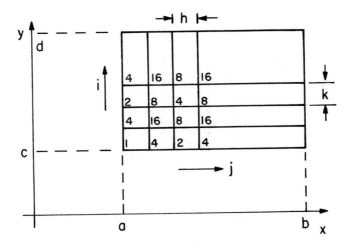

Figure 7.5. Simpson's rule for a rectangle.

Because of the properties of Simpson's rule with respect to integration in either the *x*- or the *y*-direction, Equation (7.26) will be exact if, within each subrectangle of sides $(2h, 2k)$, $f(x, y)$ is a linear combination of terms of the form $x^p y^q$, where $p \leq 3, q \leq 3$.

It is clear that other integration formulas (including Gaussian) can be used similarly. If there are *n* mesh points in each direction, then the number of functional evaluations required is n^2 in the two-dimensional case; in *N* dimensions, it is n^N. For large *N* the size of this number may induce one to return to the Monte Carlo method of Chapter 4.

If the region of integration *A* is not rectangular, then one approach would be to imbed the region in a rectangle, with *f* defined to vanish outside of *A*. This has the disadvantage that *f* is now not well fitted by linear combinations of products of powers in *x* and *y*, so that rules like Simpson's tend to lose their accuracy. It is, of course, also possible to integrate over the actual region, first in one direction and then in the other, using appropriate sets of mesh points for each individual integration. In Figure 7.6, for example, each of the *y*-integrations could use Simpson's rule; each choice of *x* would require a different set of mesh points for the *y*-integration.

It is also possible to transform a given region *A* into a rectangle by an appropriate change in variables. Sometimes this is easy; for example, if *A* is the interior of a circle, polar coordinates could be used.

For an appropriate region *A*, it may be useful to try to choose evaluation points $(x_1, y_1), \cdots, (x_n, y_n)$ within the region and corresponding coefficients $w_1, \cdots, w_n$ so that

$$\int_A f(x, y) \, dx = w_1 f(x_1, y_1) + \cdots + w_n f(x_n, y_n)$$

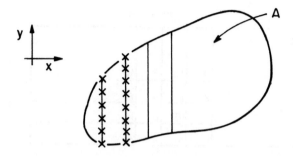

Figure 7.6. Integration over A, with respect to y, then with respect to x.

is exact for all functions f of the form $x^p y^q$, up to some maximum values of p and of q. This represents an extension of the idea of Gaussian quadrature. Some results of this character will be found in books by Abramowitz and Stegun (1964) and Davis and Rabinowitz (1984).

7.6 FAST FOURIER TRANSFORM

It is often desirable to approximate a periodic function by a finite sum of harmonic functions. This situation occurs in the analysis of tides, in the determination of the output of an electric filter resulting from a given input signal, in seismic analysis, in the evaluation of possible resonances in a mechanical system, in the detection of cycles of economic activity, and in many other areas.

Let $f(x)$ be periodic, with a period 2π (if $f(x)$ has any other period, the period can be made equal to 2π by a change in variable). Suppose that it is desired to approximate $f(x)$ with the interval $(-\pi, \pi)$ by a sum of sines and cosines of the form

$$f(x) \cong \frac{a_0}{2} + \sum_{j=1}^{n} (a_j \cos jx + b_j \sin jx) , \qquad (7.27)$$

where the a_j and b_j are constants and n is some fixed integer. The initial constant is written as $a_0/2$ rather than as a_0 for future convenience. A reasonable way in which to choose the a_j and b_j is so that the mean square discrepancy between the two sides of Equation (7.27) over the interval $(-\pi, \pi)$ should be minimal. If this is done, then the partial derivative of I with respect to each a_j and each b_j must vanish when these coefficients have their optimal values. Here

$$I = \int_{-\pi}^{\pi} \left[f(x) - \left\{ \frac{1}{2} a_0 + \sum_{j=1}^{n} (a_j \cos jx + b_j \sin jx) \right\} \right]^2 dx .$$

Using the easily proved facts

$$\int_{-\pi}^{\pi} \cos px \cos qx \, dx = \begin{cases} 0, & p \neq q, \\ \pi, & p = q \neq 0, \\ 2\pi, & p = q = 0, \end{cases}$$

$$\int_{-\pi}^{\pi} \sin px \cos qx \, dx = 0,$$

$$\int_{-\pi}^{\pi} \sin px \sin qx \, dx = \begin{cases} 0, & p \neq q, \\ \pi, & p = q \neq 0, \end{cases}$$

where p and q are nonnegative integers, the differentiation process yields

$$a_j = \frac{1}{\pi} \int_{-\pi}^{\pi} f(x) \cos jx \, dx, \qquad j = 0, 1, 2, \ldots,$$

$$b_j = \frac{1}{\pi} \int_{-\pi}^{\pi} f(x) \sin jx \, dx, \qquad j = 1, 2, \ldots. \tag{7.28}$$

[The fact that the same formula now holds for a_0 as for a_1, a_2, ... motivates the factor $\frac{1}{2}$ in Eq. (7.27).]

Observe that the formulas for the a_j and b_j given in Equations (7.28) do not involve n, the number of terms used in the "trigonometric polynomial" of Equation (7.27). Thus, if it is desired to include higher-frequency terms in Equation (7.27), the previously obtained coefficients do not have to be recalculated. Equation (7.27), with the a_j and b_j defined by Equations (7.28), represents the initial portion of a *Fourier series*. It can be shown that as $n \to \infty$ the resulting series will converge to $f(x)$ in $(-\pi, \pi)$ if $f(x)$ satisfies rather mild conditions; however, we will restrict out interest to the case of finite n.

It has already been remarked in Section 7.2 that the trapezoidal rule is appropriate for the integration of periodic functions. Choose mesh points x_k in the interval $(-\pi, \pi)$ defined by $x_k = k\pi/N$ for $k = -N, -N + 1, \ldots, 0, \ldots, N - 1, N$, where N is some chosen positive integer. Denote the value of $f(x_k)$ by f_k, noting that, because of periodicity, $f_{-N} = f_N$. Then the trapezoidal rule applied to Equations (7.28) gives

$$a_j = \frac{1}{N} \sum_{k=-N}^{N-1} f_k \cos \frac{jk\pi}{N},$$

$$b_j = \frac{1}{N} \sum_{k=-N}^{N-1} f_k \sin \frac{jk\pi}{N}. \tag{7.29}$$

It is customary to make n and N about the same (cf. Prob. 7.17). Each summation in Equations (7.29) requires $2N$ operations (an operation is here a multiplication and an addition—the trigonometric terms can be stored), and if we need N of the a_j and b_j quantities, the total operation count is of order $4N^2$. This can easily become excessive, particularly if this kind of spectral decomposition is done repetitively. The *Fast Fourier Transform* is a technique that permits the calculation of the a_j and b_j in a more economical manner, provided that N can be written as $N = r_1 r_2$, where r_1 and r_2 are integers greater than unity.

The discussion becomes clearer if Equations (7.29) are first written in terms of complex numbers. Write

$$c_j = a_j + ib_j = \frac{1}{N} \sum_{k=-N}^{N-1} f_k\, e^{ijk\pi/N} ,$$

where, as usual, $i = \sqrt{-1}$. Define $z = e^{i\pi/N}$ so that

$$c_j = \frac{1}{N} \sum_{k=-N}^{N-1} f_k z^{jk} . \tag{7.30}$$

Assuming as above that N is composite of the form $N = r_1 r_2$, write

$$k = \alpha r_1 + \beta , \qquad j = 2\gamma r_2 + \delta ,$$

where α, β, γ, δ are integers. Then the summation over k can be achieved by two individual summations: over α from $-r_2$ to $r_2 - 1$, and over β from 0 to $r_1 - 1$ (for each requisite term will appear in the new summation). Thus,

$$c_j = \frac{1}{N} \sum_{\beta=0}^{r_1-1} \sum_{\alpha=-r_2}^{r_2-1} f_{\alpha r_1 + \beta} z^{(\alpha r_1 + \beta)j} . \tag{7.31}$$

But

$$z^{(\alpha r_1 + \beta)j} = z^{\alpha r_1 j} \cdot z^{\beta j} = z^{\alpha r_1 (2\gamma r_2 + \delta)} \cdot z^{\beta j} = z^{\alpha \delta r_1} \cdot z^{\beta j} ,$$

since $z^{2\alpha\gamma\, r_1 r_2} = (z^{2N})^{\alpha\gamma} = 1$. So Equation (7.31) becomes

$$c_j = \frac{1}{N} \sum_{\beta=0}^{r_1-1} \left(\sum_{\alpha=-r_2}^{r_2-1} f_{\alpha r_1 + \beta} z^{\alpha r_1 \delta} \right) z^{\beta j} . \tag{7.32}$$

The quantity in parentheses has to be evaluated for each of r_1 values of β and each of $2r_2$ values of δ (i.e., for $2r_1 r_2 = 2N$ combinations), and each evaluation requires $2r_2$ operations. The operation count for this process is therefore $(2r_2)(2N) = 4Nr_2$. With the quantities in the parentheses tabulated, each summation over β requires r_1 operations; this must be done for N values of j, giving an operation count of Nr_1. The total is $N(4r_2 + r_1)$, which is usually much less than $2N^2$. (For example, with $N = 100 = 10 \times 10$, $2N^2 = 20{,}000$ and $N(4r_2 + r_1) = 5{,}000$.) If either r_1 or r_2 can be factored, an analogous further improvement is possible. Of course, a similar process may be used to reconstitute the mesh point values for any given c_j values.

ANNOTATED BIBLIOGRAPHY

M. Abramowitz and I. Stegun, eds., 1964, *Handbook of Mathematical Functions*, National
Bureau of Standards, Applied Mathematics Series 55, U.S. Government Printing Office,
Washington, D.C., 1046p. (Dover reprint available.)

Section 25 gives an extensive collection of Newton–Cotes formulas, weights, and
abscissae for Gaussian integration (see Table 25.4) and related formulas.

F. S. Acton, 1970 *Numerical Methods that Work*, Haper & Row, New York, 539p.

Chapter 4, dealing with quadrature, is one of the best chapters of this very readable
text. The treatment of "tails" and of singularities is particularly worthwhile.

P. J. Davis and P. Rabinowitz, 1984 *Methods of Numerical Integration*, 2nd ed., Academic
Press, New York, 612p.

This survey of the current state of the art in numerical integration includes a discussion
of multiple integration, infinite integrals, and oscillatory integrands.

PROBLEMS

7.1 Verify the second equation of (7.7) and the second equation of (7.8).
Use all equations of (7.2), (7.7), and (7.8) to experiment with the numerical
evaluation of the derivatives of $f(x) = 1/x$, where f is taken as known at each of
three equally spaced points. Are the error estimates verified? How well does Richardson extrapolation work?

7.2 a: Verify the last equation of (7.9). b: The first and last equations of
(7.7) do not appear symmetrical, at least superficially. Thus, we do not get one
from the other by simply interchanging the subscripts 1 and 3. Why not?

7.3 Suppose that an error of size ϵh, where $|\epsilon| \ll 1$, is made in the deter-
mination of f_1, f_2, or f_3. What errors are made in f_0' and f_0'' by using Equations
(7.7) and (7.8)?

7.4 Verify the error estimate made for the integral J following Equation
(7.13). [Hint: Differentiate the integrand twice and obtain a bound for each term
by giving the numerator its largest and the denominator its smallest absolute values
attained anywhere in the interval.]

7.5 Derive Equation (7.14). [*Hint*: For interpolation points at $-h$, $-\epsilon$, ϵ,
h, where $0 < \epsilon \ll 1$, use Newton's formula to write

$$f(x) = f_{-h} + (x + h) [\ ] + \cdots + (x^2 - h^2)(x - \epsilon) [\ ]$$

$$+ \frac{1}{4!} (x^2 - h^2)(x^2 - \epsilon^2) F^{(4)}(\xi) ,$$

and let $\epsilon \to 0$; the second to the last term is odd and makes no contribution.]

7.6 The error function of Chapter 4 is defined by

$$\operatorname{erf}(x) = \frac{2}{\sqrt{\pi}} \int_0^x e^{-t^2} \, dt .$$

Use Newton–Cotes and Gaussian integration methods to evaluate erf(1) with an error of less than 10^{-9}. (*Ans:* .84270 07929) Which method is better?

7.7 The modified Bessel function of the first kind of order n (n integral) may be defined by

$$I_n(x) = \frac{1}{\pi} \int_0^\pi e^{x \cos \theta} \cos n\theta \, d\theta .$$

Find $I_3(7)$ to eight significant figures. For what range of x and n values would numerical integration be a practical way to evaluate $I_n(x)$, say as a library subroutine? (Assume a few hundred mathematical operations as an upper limit. Many other algorithms for Bessel functions exist.)

7.8 Integration by parts of $I = \int_0^h f(x) \, dx$ gives

$$I = \left[f(x) \cdot \left(x - \frac{h}{2} \right) \right]_0^h - \int_0^h \left(x - \frac{h}{2} \right) f'(x) \, dx$$

$$= \frac{h}{2} [f(0) + f(h)]$$

$$- \left\{ f'(x) \left[\frac{1}{2} \left(x - \frac{h}{2} \right)^2 + k_1 h^2 \right]_0^h - \int_0^h \left[\frac{1}{2} \left(x - \frac{h}{2} \right)^2 + k_1 h^2 \right] f''(x) \, dx \right\}$$

$$= \cdots .$$

Continue in this way, choosing constants $k_1, k_2, \ldots$ appropriately, to obtain

$$I = \frac{h}{2} [f(0) + f(h)] + \alpha h^2 [f'(h) - f'(0)] + \beta h^4 [f'''(h) - f'''(0)] + \cdots .$$

Add these results over all subintervals to obtain

$$I = \text{trapezoidal rule} + ah^2 + bh^4 + \cdots ,$$

noting that the coefficients $a, b, \ldots$ involve the difference of the first, third, $\ldots$, derivatives at the endpoints. The result of this problem is used in the discussion of Romberg integration in Section 7.2. An elaboration of this process leads to the *Euler–Maclaurin sum formula*, which is useful in providing an end correction to the trapezoidal formula, in asymptotics, and in summing series.

7.9 Write an adaptive mesh subroutine, based on Simpson's rule, that integrates a given function $f(x)$ to a desired accuracy. Assume $f(x)$ is well behaved. Permit the user to prescribe an upper limit on the number of steps.

7.10 One way of defining the Legendre polynomial of order n, within a multiplicative constant, is as the polynomial solution of Legendre's differential equation.

$$(1 - x^2)y'' - 2xy' + n(n + 1)y = 0 .$$

By substituting $y = a_0 + a_1x + \cdots + a_nx^n$ into this equation, obtain the recursion relation

$$(j + 2)(j + 1) a_{j+2} = [j(j + 1) - n(n + 1)]a_j ,$$

and verify that a solution having the form of an nth-degree polynomial does exist.

Next, write the differential equation satisfied by $P_m(x)$ in the form

$$[(1 - x^2)P'_m]' + m(m + 1)P_m = 0$$

and multiply each term by P_n, where $n \neq m$. Repeat, interchanging the roles of m and n. Subtract and integrate the result between -1 and 1. Using integration by parts, deduce that

$$\int_{-1}^{1} P_m(x)P_n(x) \, dx = 0 \qquad \text{for } m \neq n ,$$

so that the P_n are indeed orthogonal over $(-1, 1)$.

So far, each P_n as defined above still contains an arbitrary multiplicative constant. The standard choice for this constant is such that

$$\int_{-1}^{1} P_n^2(x) \, dx = \frac{2}{2n + 1}.$$

Making this choice, show that the first few $P_n(x)$ are given by

$$P_0 = 1, \quad P_1 = x, \quad P_2 = \tfrac{3}{2}x^2 - \tfrac{1}{2}, \quad P_3 = \tfrac{5}{2}x^3 - \tfrac{3}{2}x,$$

$$P_4 = \tfrac{1}{8}(35x^4 - 30x^2 + 3), \quad P_5 = \tfrac{1}{8}(63x^5 - 70x^3 + 15x.$$

It looks as if $P_n(x)$ is an odd function of x for odd n, and an even function for even n. Is this true in general?

7.11 *a*: Evaluate

$$\int_0^1 [(e^t - 1)/t] \, dt$$

to eight significant figures, using a Gaussian quadrature method. *b*: Repeat for

$$\int_0^1 \exp(t^{2/3}) \, dt .$$

c: Repeat for

$$\int_{-1}^{1} \frac{\cos x}{\sqrt{1 - x^2}} \, dx \ .$$

[*Hint*: For *a* use as interpolation points the zeros of Legendre polynomials. For *c* use both Chebyshev polynomials and Legendre polynomials and compare results. For *b* you are on your own. Note that your results for *a* and *b* are easily checked by expansion of the integrand in a power series, followed by term-by-term integration.]

7.12 Let

$$I(x) = \int_{0}^{1} \frac{e^{-t^2}}{t + x} \, dt \qquad \text{for } x > 0 \ .$$

Use differentiation with respect to *x* and integration by parts to show that

$$I'(x) + 2xI = \frac{e^{-1}}{1 + x} - \frac{1}{x} + 2 \int_{0}^{1} e^{-t^2} \, dt \ .$$

Evaluate $I(1)$ by any method, and then solve this differential equation numerically to find $I(1 + .1n)$ for $n = 1, 2, \ldots , 10$. Six significant figures are required.

7.13 Write a subroutine based on Gaussian quadrature that uses four mesh points per subinterval, and use it to evaluate

$$a: \int_{0}^{3} \sin(x + \sqrt{x}) \, dx \qquad b: \int_{0}^{1} \frac{dt}{1 - \sqrt{t} + t^2}$$

$$c: \int_{0}^{1} e^{-3 \cos \theta} \, d\theta \qquad d: \int_{1.01}^{\pi} \frac{e^{-x^2} - 1}{x - 1} \, dx \ .$$

Six significant figures are required. What test cases did you run?

7.14 Use the method of Section 7.4, as applied there to erfc(*x*), to evaluate $I = \int_{0}^{\infty} e^{-x \cosh \theta}$ for large positive *x*. At $x = 1$ and $x = 10$ how well does the asymptotic result agree with that obtained by numerical integration? How large should *x* be for adequate agreement?

7.15 The *gamma function*, $\Gamma(x)$, is defined by

$$\Gamma(x) = \int_{0}^{\infty} e^{-t} t^{x - 1} \, dt$$

for $x > 0$. Write a subroutine that generates the value of $\Gamma(x)$ for any *x* satisfying $0 < x < 10$ [*Hint*: The recursion relation $\Gamma(x + 1) = x\Gamma(x)$, proved by integration by parts, is useful.] How many significant figures, or what error bound, can you guarantee?

7.16 Evaluate

$$\int_0^\infty \frac{e^{-x}}{1 + \sqrt{x} + x^2} \, dx$$

to at least six significant figures. Repeat with the numerator of the integrand replaced by $e^{-x} \sin x$.

7.17 A student claims that if $n = N$ in Equation (7.27), and if a_j, b_j are defined by Equations (7.29), then,

$$f(x_k) = \frac{a_0}{2} + \sum_{j=1}^{N} (a_j \cos jx_k + b_j \sin jx_k)$$

exactly for each value of k, where $x_k = k\pi/N$. Is he right?

7.18 Write a subroutine that uses Simpson's rule to integrate over a rectangle. Experiment with the function $f(x, y) = 1/(x^2 + y^2 + 1)$ for $x^2 + y^2 \leq 1$, $f(x, y) = 0$ otherwise, in the rectangular region $0 < x < 1$, $0 < y < 1$.

7.19 A black body radiates energy according to Planck's formula

$$e \, d\lambda = \frac{2\pi hc^2 \, d\lambda}{\lambda^5 [\exp (hc/k\lambda T) - 1]} \frac{\text{ergs}}{\text{cm}^2 \text{ sec}},$$

where $e \, d\lambda$ is the energy radiated within a wavelength interval of $d\lambda$ cm, λ is the wavelength in cm, h is Planck's constant (6.6256×10^{-27} erg sec), c is the speed of light (2.99792×10^{10} cm/sec), k is Boltzmann's constant (1.3805×10^{-16} erg/°K), and T is the absolute temperature in °K. Compute the energy emitted in each of the wavelength intervals (0, 10) (100, 110), (1000, 1010), (0, ∞). Consider the cases $T = 10, 100, 1000$.

7.20 Find the solid angle subtended at the origin by a circular disk of radius 1 unit, whose center is on the *x*-axis at a distance of three units from the origin, and whose normal makes an angle of 52° with the positive *x*-axis. What checks did you use? Problems of this kind often arise in optics.

PARTIAL DIFFERENTIAL EQUATIONS

8.1 A DIFFUSION PROBLEM

Equations of diffusion type are encountered in problems of neutron or gaseous diffusion, in thermal conduction, in ground water seepage, and in many other areas. For definiteness we consider here a thermal problem.

Suppose that heat is flowing through a thick wall of uniform thickness in such a way that the temperature ϕ depends only on the distance x from one face of the wall and on the time t (see Fig. 8.1a). This situation can be achieved, to a close degree of approximation, if the extent of the wall in the y- and z-directions is very large, and if the initial conditions and wall boundary temperatures are independent of the y- and z-coordinates.

Consider now an elemental portion dV of the material inside the wall, of thickness dx and of unit area perpendicular to the x-axis, as shown in Figure 8.1b. The thermal conductivity k of a medium is defined to be the ratio of the heat flow per unit area to the thermal gradient, and, consequently (since we have unit area), the rate of heat flow across the face at position x equals $[k(\partial\phi/\partial x)]|_x$, where the subscript means that the expression in brackets is to be evaluated at position x. At the back face the rate of heat flow is, similarly, $[k(\partial\phi/\partial x)]|_{x+dx}$, where the evaluation is now for the face located at $x + dx$. The difference between these two quantities is given by

$$\left(k\,\frac{\partial\phi}{\partial x} \right)\Bigg|_{x+dx} - \left(k\,\frac{\partial\phi}{\partial x} \right)\Bigg|_{x} = \frac{\partial}{\partial x}\left(k\,\frac{\partial\phi}{\partial x} \right)\cdot dx \;, \qquad \textbf{(8.1)}$$

where we have used the definition of a partial derivative. This expression represents the net rate of heat accumulation inside dV. If k were a constant, it could be taken

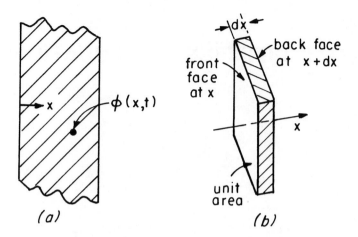

Figure 8.1. *a:* Portion of wall. *b:* Volume element.

outside the $\partial/\partial x$ symbol in Equation (8.1); however, we will permit k to be a function of x, denoted by $k(x)$.

Next, define the thermal capacity c of the material to be the amount of heat required to raise the temperature of a unit volume by one degree. Then the rate of heat accumulation inside dV is given by

$$(c)(1)(dx)\,\frac{\partial \phi}{\partial t}\,, \tag{8.2}$$

since the volume of dV is $(1)(dx)$. In general, c may depend on both x and ϕ, and we will denote it by $c(x, \phi)$.

Thus, we have obtained two different expressions, (8.1) and (8.2), for the rate of heat accumulation within dV, and they must, of course, be equal. It follows that

$$c(x, \phi)\,\frac{\partial \phi}{\partial t} = \frac{\partial}{\partial x}\left[k(x)\,\frac{\partial \phi}{\partial x}\right] \tag{8.3}$$

is the partial differential equation that must be satisfied by $\phi(x, t)$. On physical grounds we would expect to also have to specify the initial temperature distribution, $\phi(x, 0)$, and the temperatures of the two faces, say $\phi(0, t)$ and $\phi(l, t)$, if the faces are at $x = 0$ and at $x = l$, in order to completely characterize the problem.

Before turning to numerical methods, it is worthwhile to look for exact solutions of Equation (8.3) for some special cases; such solutions will throw some light on the way in which $\phi(x, t)$ can be expected to behave and will also serve as test cases for numerical methods.

Suppose that k and c are constant, and denote the value of k/c by α^2 (called the *diffusivity*). Then Equation (8.3) takes the form

$$\frac{\partial \phi}{\partial t} = \alpha^2 \frac{\partial^2 \phi}{\partial x^2} . \tag{8.4}$$

Try the possible solution

$$\phi(x, t) = \Omega(t) \sin \frac{n\pi x}{l} , \tag{8.5}$$

where n is a positive integer. Substitution into Equation (8.4) yields

$$\Omega' = -\alpha^2 \left(\frac{n\pi}{l}\right)^2 \Omega , \tag{8.6}$$

so that the expression (8.5) is indeed a solution, provided that $\Omega(t)$ satisfies the ordinary differential equation (8.6). The solution of Equation (8.6) is

$$\Omega = A e^{-(n\pi\alpha/l)^2 t} ,$$

where A is any constant. Thus,

$$\phi(x, t) = A e^{-(n\pi\alpha/l)^2 t} \sin \frac{n\pi x}{l} \tag{8.7}$$

is a special solution of Equation (8.4). At $t = 0$ it has the form $\phi(x, 0) = A \sin (n\pi x/l)$, and at $x = 0$, $x = l$, it vanishes. Consequently, the function $\phi(x, t)$ defined by Equation (8.7) is a solution of the heat diffusion equation (8.4) that corresponds to an initial sinusoidal temperature distribution inside the wall and to the case in which the two faces of the wall are kept at $0°$ ("0" can be any reference temperature with respect to which ϕ is measured).

The special solution (8.7) has the interesting feature that it decays with time and, moreover, more rapidly so as n increases, as α increases, or as l decreases. This is certainly in accordance with physical expectation. For example, as n increases, the initial temperature distribution becomes more oscillatory, and the associated initial temperature gradients become greater, thus inducing more rapid heat flow. Similarly, an increase in α can be attained by increasing k or by decreasing c, and either of these changes will tend to make thermal equilibrium be attained more rapidly. Finally, a decrease in l increases the initial temperature gradient, so this effect is also as expected.

Special solutions of the form (8.7) can be superposed because of the linearity of Equation (8.4) (for the case $\alpha =$ constant). Let

$$\phi(x, 0) = a_1 \sin \frac{\pi x}{l} + a_2 \sin \frac{2\pi x}{l} + \cdots + a_N \sin \frac{N\pi x}{l} . \tag{8.8}$$

Then it follows from the above that

$$\phi(x, t) = a_1 e^{-(\pi\alpha/l)^2 t} \sin \frac{\pi x}{l} + a_2 e^{-(2\pi\alpha/l)^2 t} \sin \frac{2\pi x}{l}$$

$$+ \cdots + a_N e^{-(N\pi\alpha/l)^2 t} \sin \frac{N\pi x}{l} \tag{8.9}$$

is a solution of Equation (8.4) corresponding to the initial condition (8.8). (The physical problem certainly has a unique solution; if the differential equation (8.4) as well as the initial and boundary conditions adequately describe the physical problem, then the solution (8.9) can also be expected to be unique. Uniqueness can also be proved mathematically, but here we simply accept it on physical grounds.) The righthand side of Equation (8.8) represents the first few terms of a *Fourier sine series*, which has the capability, as N increases indefinitely, of representing any initial temperature distribution vanishing at the two ends (and satisfying rather mild well-behavedness conditions). Again, we merely note this fact as a matter of interest and leave the detailed discussion to texts on Fourier series.

To give some idea as to the time scale associated with the solution (8.7), typical values of α^2 for water, rock, and copper are .0015, .01, and 1.1 cm^2/sec, respectively. Thus, for a copper sheet of 1-cm thickness, an initial sin πx temperature distribution decays by half in about .058 seconds; if the material were rock, it would take 702 seconds.

Returning now to the general situation described by Equation (8.3), with $c = c(x, \phi)$ and $k = k(x)$, and permitting $\phi(x, 0)$, $\phi(0, t)$, and $\phi(l, t)$ to be specified rather arbitrarily, one can hardly expect to be able to find analytical solutions, and it is necessary to consider numerical methods.

8.2 A NUMERICAL METHOD FOR THE DIFFUSION EQUATION

Although our interest is in the more general equation (8.3), we begin with Equation (8.4), with α^2 constant, in order to explore the properties of numerical techniques in an environment that is as simple as possible. Divide the region (0, l) into N subintervals of length $\Delta x = l/N$, and consider also time intervals of length Δt. Denote the numerical approximation to $\phi(j\Delta x, k\Delta t)$ by ϕ_j^k. The array of mesh points in space and time is depicted in Figure 8.2. From Section 7.1 it is known that a suitable approximation for $\partial^2\phi/\partial x^2$ (which is simply a second derivative in the x-direction, holding t fixed) at the mesh point (j, k) is

$$\frac{\partial^2\phi}{\partial x^2}\bigg|_{j,k} = \frac{\phi_{j+1}^k - 2\phi_j^k + \phi_{j-1}^k}{\Delta x^2}. \tag{8.10}$$

Also, a suitable approximation to $\partial\phi/\partial t$ at the same mesh point is

$$\frac{\partial\phi}{\partial t}\bigg|_{j,k} = \frac{\phi_j^{k+1} - \phi_j^k}{\Delta t}. \tag{8.11}$$

Using these two approximations, we replace Equation (8.4) by the approximation

$$\phi_j^{k+1} - \phi_j^k = \beta(\phi_{j+1}^k - 2\phi_j^k + \phi_{j-1}^k) \tag{8.12}$$

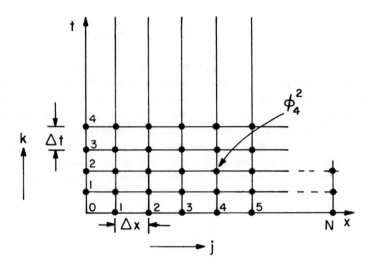

Figure 8.2. Mesh points for diffusion (or wave) equation.

for $j = 1, 2, \ldots, N - 1$, where $\beta = \alpha^2 \Delta t/(\Delta x)^2$. Suppose that the end temperatures are specified for all time, so that ϕ_0^k and ϕ_N^k are known for all k. Assume also that the value of ϕ_j^0 is known from the initial conditions (we set $\phi_j^0 = \phi(j\Delta x, 0)$ for $j = 0, 1, 2, \ldots, N$). Equation (8.12) then permits the calculation of all the ϕ_j^1 values, then of all the ϕ_j^2 values, and so on, in what is called a "time-stepping" process. It would seem that the numerical solution of Equation (8.4) is straightforward, and that it is only necessary to choose small enough values of Δt and of Δx that the action can be followed with reasonable fidelity.

Unfortunately (as numerical analysts discovered long ago), this process can go badly astray, and this can be shown even without numerical experimentation by again considering a special sinusoidal case. To avoid having to satisfy end conditions at $x = 0$ and at $x = l$, let the wall be infinitely thick, stretching from $x = -\infty$ to $x = +\infty$, and let $\phi(x, 0) = A \sin \lambda x$, where λ is any constant. This particular initial condition can be thought of as one component of a more general initial temperature distribution. Then

$$\phi_j^0 = A \sin \lambda j \Delta x \, ,$$

and, from Equation (8.12) by direct calculation

$$\phi_j^1 = A(1 - 2\beta + 2\beta \cos \lambda \Delta x) \sin \lambda j \Delta x$$

$$= A\left(1 - 4\beta \sin^2 \frac{\lambda \Delta x}{2}\right) \sin \lambda j \Delta x \, .$$

Continuing, we find

$$\phi_j^k = A \left(1 - 4\beta \sin^2 \frac{\lambda \Delta x}{2}\right)^k \sin \lambda j \Delta x . \tag{8.13}$$

Now λ can have almost any value, as we experiment with different initial wave-lengths, and so the quantity in parentheses can be expected to have almost any value between 1 and $1 - 4\beta$. But if $\beta > \frac{1}{2}$, then $1 - 4\beta < -1$, so $|1 - 4\beta| > 1$, and, for some appropriate value of λ, the coefficient of the sinusoidal term in Equation (8.13) will grow as k increases. In fact, the growth is essentially exponential, so that any component in the original temperature profile that corresponds to this particular λ will be magnified enormously as time passes. Since computer roundoff error will always introduce components of various wavelengths, the result is bound to be disastrous.

Thus, we must have

$$\beta = \frac{\alpha^2 \Delta t}{\Delta x^2} < \frac{1}{2} \tag{8.14}$$

for stability. Next suppose that condition (8.14) is satisfied; using Equation (8.12), can we expect the numerical solution to approach the exact solution of Equation (8.4) as Δt and Δx become small? Consider now a particular sinusoidal component for the case in which ϕ vanishes at the two ends, that is, at $x = 0, l$. Let

$$\phi(x, 0) = A \sin \frac{n \pi x}{l} ,$$

so that, from Equation (8.7),

$$\phi(x, t) = A e^{-(n \pi \alpha / l)^2 t} \sin \frac{n \pi x}{l} . \tag{8.15}$$

Using Equation (8.13), the corresponding numerical solution (with $\lambda = n\pi/l$) is

$$\phi_j^k = A \left(1 - 4\beta \sin^2 \frac{n \pi \Delta x}{2l}\right)^k \sin \frac{n \pi j \Delta x}{l} . \tag{8.16}$$

With $j \Delta x = xj$, the sinusoidal terms in Equations (8.15) and (8.16) agree, so it is only necessary to look at the coefficients. Now for small Δx and small Δt,

$$\left(1 - 4\beta \sin^2 \frac{n \pi \Delta x}{2l}\right)^k \cong \left[1 - 4\beta \left(\frac{n \pi \Delta x}{2l}\right)^2 + \cdots \right]^k$$

$$\cong \left[1 - \frac{\alpha^2 n^2 \pi^2 \Delta t}{l^2} + \cdots \right]^k$$

$$\cong \exp \left[k \ln \left(1 - \frac{\alpha^2 n^2 \pi^2 \Delta t}{l^2} + \cdots \right)\right]$$

$$\cong \exp\left[-k\,\frac{\alpha^2 n^2 \pi^2 \Delta t}{l^2} + \cdots\right]$$

$$\cong \exp\left[-\frac{\alpha^2 n^2 \pi^2}{l^2}\,t + \cdots\right].$$

Thus, the leading term does indeed correspond to the result of Equation (8.15), and we can expect convergence, since the higher-order terms become negligible as Δt and $\Delta x \to 0$ (of course, Equation (8.14) must always be satisfied).

The above expectation is borne out by computational experience. If the stability criterion (8.14) is satisfied, and if Δx, Δt have been chosen small enough in comparison with the scales of distance and time with which $\phi(x, t)$ changes, then excellent results can be achieved by using Equation (8.12). It may be remarked that condition (8.14) can be onerous if very high accuracy is desired, since each decrease in Δx by factor of 2 requires that Δt be decreased by a factor of 4.

In the more general case in which the coefficient of α^2 in Equation (8.4) depends on x or on ϕ, it is customary to require Equation (8.14) to be satisfied for the largest value of α^2 experienced anywhere in the interval.

Equation (8.12) uses an accurate approximation for the second derivative, but a less accurate one for the first derivative. One might be tempted to replace it by

$$\frac{\phi_j^{k+1} - \phi_j^{k-1}}{2\,\Delta t} = \alpha^2\,\frac{\phi_{j+1}^k - 2\phi_j^k + \phi_{j-1}^k}{(\Delta x)^2} \tag{8.17}$$

in which a centered expression is used for $\partial\phi/\partial t$; however, an analysis similar to the preceding one shows that this new equation is always unstable. It is better to take advantage of the stability associated with "feedback," as mentioned in connection with ordinary differential equations in Chapter 6, and use an "implicit" equation:

$$\frac{\phi_j^{k+1} - \phi_j^k}{\Delta t} = \frac{\alpha^2}{2}\left[\frac{\phi_{j+1}^k - 2\phi_j^k + \phi_{j-1}^k}{(\Delta x)^2} + \frac{\phi_{j+1}^{k+1} - 2\phi_j^{k+1} + \phi_{j-1}^{k+1}}{(\Delta x)^2}\right]. \tag{8.18}$$

Note that the second derivative term is now averaged over the time intervals k and $k + 1$. Equation (8.18), often referred to as the *Crank–Nicholson* method, is stable for all Δt and Δx; however, the constraint that Δt and Δx must be small enough to follow the motion accurately still remains.

Equation (8.18) is implicit rather than explicit, in the sense that a set of algebraic equations must be solved at each time step. However, the coefficient matrix is tridiagonal, so (cf. Prob. 2.13) Gaussian elimination is very fast.

The boundary conditions may be more complicated than those considered so far. If the wall is insulated at $x = l$, then $\partial\phi/\partial x = 0$ there. One way of handling this condition is to insist that $\phi_{N-1}^k = \phi_N^k$ for all k (where $x_N = l$); this identity is used to replace ϕ_N^k by ϕ_{N-1}^k for the case $j = N - 1$ in Equation (8.12). More accurately (cf. Sec. 7.1) we could write $3\phi_N^k = 4\phi_{N-1}^k - \phi_{N-2}^k$. In the case of Equation (8.18), a similar replacement would be made for ϕ_N^{k+1}. More complicated "radiation-type" boundary conditions, in which a linear relation between end values of ϕ and $\partial\phi/\partial x$ is imposed, can be handled analogously.

Returning now to the general diffusion equation (8.3), an explicit numerical version would be

$$c(x_j, \phi_j^k) \frac{\phi_j^{k+1} - \phi_j^k}{\Delta t}$$

$$= \frac{1}{\Delta x}\left[k\left(\left[j + \frac{1}{2}\right]\Delta x\right)\frac{\phi_{j+1}^k - \phi_j^k}{\Delta x} - k\left(\left[j - \frac{1}{2}\right]\Delta x\right)\frac{\phi_j^k - \phi_{j-1}^k}{\Delta x}\right] \quad (8.19)$$

for $j = 1, 2, \ldots, N - 1$. This equation reduces to Equation (8.12) if c and k are constants. A similar, but more complicated (and now nonlinear) Crank–Nicholson version may be written. In these situations the stability criterion (8.14) may be only roughly correct; some experimentation may be necessary.

A two-dimensional version of Equation (8.4) is

$$\frac{\partial\phi}{\partial t} = \alpha^2\left(\frac{\partial^2\phi}{\partial x^2} + \frac{\partial^2\phi}{\partial y^2}\right), \quad (8.20)$$

where ϕ is now a function of x, y, and of time t. This equation would govern the flow of heat along a thin plate lying in the (x, y) plane, whose upper and lower faces are insulated. Let R denote the region in the (x, y) plane corresponding to the plate, and cover R by a square net of mesh points as shown in Figure 8.3. A curved boundary Γ may be replaced by a jagged boundary, so that the revised region (shown shaded in Fig. 8.3) is bordered by mesh lines. If the temperature, as a function of time, is specified along the original curved boundary, then the values of ϕ at the mesh points A lying on the revised border are obtained by

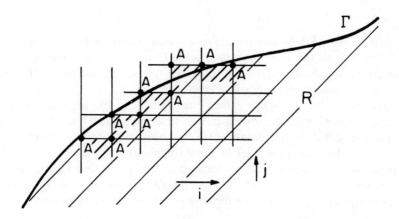

Figure 8.3. Mesh points near boundary.

interpolation (cf. Prob. 8.6). Denote the approximation to $\phi(x, y, t)$, at the mesh point (i, j) and at time $(k \Delta t)$ (see Fig. 8.3) by $\phi_{i,j}^k$. Then if i, j is an interior point, the equivalent of Equation (8.12) becomes

$$\frac{\phi_{i,j}^{k+1} - \phi_{i,j}^k}{\Delta t}$$
$$= \alpha^2 \left[\frac{\phi_{i+1,j}^k - 2\phi_{i,j}^k + \phi_{i-1,j}^k}{(\Delta x)^2} + \frac{\phi_{i,j+1}^k - 2\phi_{i,j}^k + \phi_{i,j-1}^k}{(\Delta y)^2} \right]. \quad (8.21)$$

Usually, Δx and Δy are made equal, say to h. The stability criterion (8.14) then becomes

$$\frac{\alpha^2 \Delta t}{h^2} < \frac{1}{4}. \quad (8.22)$$

In two dimensions the analogue of the implicit method [Eq. (8.18)] consists in averaging the righthand side of Equation (8.21) over the two time values k and $k + 1$. The coefficient matrix is no longer tridiagonal, so the solution process for each time step is more difficult. An alternative method, due to Peaceman, Rachford, and Douglas, is to use different formulas for alternate time steps. In each pair of time steps the first is calculated by replacing k by $k + 1$ in the first term (only) on the righthand side of Equation (8.21), and the second by making this replacement in the second term (only). This stable method is referred to as the *alternating direction implicit* method or the ADI method. For an alternative method, see Problem 8.7. Similar extensions may be made to a higher number of dimensions; also, as in the one-dimensional case, the generalization to nonconstant coefficients is straightforward.

8.3 A WAVE EQUATION

Let a string be stretched between two horizontally separated points on the x-axis, say at $x = 0$ and $x = l$. Let the string be free to move in the (x, y) plane, perhaps as the result of the release of an initial deformation. Denote the horizontal component of the string tension by T, and assume the string displacement $\phi(x, t)$ is small enough that T may be taken as constant. Let the mass of the string per unit length be ρ. Then consideration of the mass acceleration in the upward direction of the string portion sketched in Figure 8.4 shows that

$$T \frac{\partial \phi}{\partial x}\bigg|_{x + dx} - T \frac{\partial \phi}{\partial x}\bigg|_x = (\rho \, dx) \frac{\partial^2 \phi}{\partial t^2} ,$$

which leads to

$$T \frac{\partial^2 \phi}{\partial x^2} = \rho \frac{\partial^2 \phi}{\partial t^2}. \quad (8.23)$$

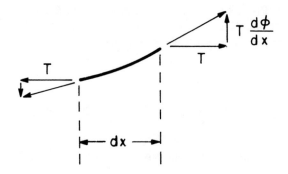

Figure 8.4. Portion of string.

We take ρ as constant. Denoting T/ρ by c^2, we can write Equation (8.23) as

$$\frac{\partial^2 \phi}{\partial t^2} = c^2 \frac{\partial^2 \phi}{\partial x^2} .$$ (8.24)

Again, it is useful to have special solutions available. One such is sinusoidal in both x and t; by direct substitution the reader can verify that

$$\phi(x, t) = A \cos \frac{n\pi c}{l} t \sin \frac{n\pi}{l} x ,$$ (8.25)

where A is a constant and n is an integer, is a solution of Equation (8.24) that vanishes for $x = 0$ and $x = l$ (and so corresponds to fixed endpoints). At $t = 0$,

$$\phi(x, 0) = A \sin \frac{n\pi x}{l} , \qquad \frac{\partial \phi}{\partial t} (x, 0) = 0 ,$$

so that the string has sinusoidal shape at $t = 0$ and is released from rest. Equation (8.25) represents a harmonic motion in which the shape of the string at any instant is just some multiple of its initial shape. As in Section 8.1, superposition could be used to provide a solution corresponding to an initial shape composed of a number of sine functions.

Actually, Equation (8.24), with constant c, is one of the few partial differential equations for which an exact general solution can be found. Introduce new variables α and β related to x and t by

$$\alpha = x + ct , \qquad \beta = x - ct .$$

Then, thinking of ϕ as being now a function of α and β and using the chain rule of differentiation, we get

$$\frac{\partial \phi}{\partial t} = \frac{\partial \phi}{\partial \alpha} \frac{\partial \alpha}{\partial t} + \frac{\partial \phi}{\partial \beta} \frac{\partial \beta}{\partial t} = \frac{\partial \phi}{\partial \alpha} c - \frac{\partial \phi}{\partial \beta} c ,$$

and, continuing this process,

$$\frac{\partial^2 \phi}{\partial t^2} = \frac{\partial^2 \phi}{\partial \alpha^2} c^2 - 2 \frac{\partial^2 \phi}{\partial \alpha \partial \beta} c^2 + \frac{\partial^2 \phi}{\partial \beta^2} c^2 .$$

Similarly,

$$\frac{\partial^2 \phi}{\partial x^2} = \frac{\partial^2 \phi}{\partial \alpha^2} + 2 \frac{\partial^2 \phi}{\partial \alpha \partial \beta} + \frac{\partial^2 \phi}{\partial \beta^2} ,$$

so that Equation (8.24) becomes

$$\frac{\partial^2 \phi}{\partial \alpha \partial \beta} = 0 . \tag{8.26}$$

Thus $(\partial/\partial \alpha)(\partial \phi/\partial \beta) = 0$, so that $\partial \phi/\partial \beta$ cannot depend on α. It follows that

$$\frac{\partial \phi}{\partial \beta} = f(\beta) ,$$

and, integrating with respect to β,

$$\phi = F(\beta) + G(\alpha)$$

(where the "constant" of integration can be some function of α, since α was held fixed during the integration). Thus, the general solution of Equation (8.24) has the form

$$\phi = F(x - ct) + G(x + ct) , \tag{8.27}$$

where F and G are arbitrary (but adequately differentiable) functions. Observe that F is a function of the combination $x - ct$ only, and so has a constant value as seen by any observer traveling to the right with velocity c. The function G maintains a fixed value for an observer traveling to the left with velocity c. For example, let the string be of infinite length, let $G = 0$, and take $F = \exp[-K(x - ct)^2]$, where K is a constant. This bell-shaped function travels (like a rigid body) to the right with velocity c. The form of the general solution (8.27) explains the name *wave equation* given to Equation (8.24).

Numerically, one could again use the mesh of Figure 8.2. Let ϕ_j^k denote the numerical approximation to $\phi(j \Delta k, k \Delta t)$; Equation (8.24) may be approximated by

$$\frac{\phi_j^{k+1} - 2\phi_j^k + \phi_j^{k-1}}{(\Delta t)^2} = c^2 \frac{\phi_{j+1}^k - 2\phi_j^k + \phi_{j-1}^k}{(\Delta x)^2} \tag{8.28}$$

for $j = 1, 2, \ldots , N - 1$ and $k = 1, 2, \ldots$. (We take $x = l$ as corresponding to $x_N = N \Delta x$.) A simple end condition is one in which ϕ is zero at each end, for all time, so that $\phi_0^k = \phi_N^k = 0$, all k. To find ϕ_j^2 by Equation (8.28), values of ϕ_j^1 and ϕ_j^0 must be known, and this is physically reasonable also, since the motion of the string will depend on its initial deformation and its initial velocity. The initial deformation specifies ϕ_j^0, and if the initial velocity at the jth mesh point is v_j, we can write

$$\frac{\phi_j^1 - \phi_j^0}{\Delta t} = v_j ,$$

so that values of ϕ_j^1 are also known. Having determined ϕ_j^0 and ϕ_j^1, values of ϕ_j^k for $k = 2, 3, \ldots$, can be obtained by repeated use of Equation (8.28).

As in the case of the diffusion equation, stability is an important consideration. Let the string have infinite length and consider a sinusoidal pattern

$$\phi_j^k = P(k) \sin \lambda j \, \Delta x , \tag{8.29}$$

where λ is a constant and $P(k)$ depends only on k. Substitution shows that for this expression to satisfy Equation (8.28) we need

$$P(k + 1) - 2P(k) + P(k - 1) = \left(\frac{c \, \Delta t}{\Delta x}\right)^2 P_k\left(-4 \sin^2 \frac{\lambda \Delta x}{2}\right)$$

or

$$P(k + 1) + P(k) \cdot \left[-2 + 4\left(\frac{c \, \Delta t}{\Delta x}\right)^2 \sin^2 \frac{\lambda \, \Delta x}{2}\right] + P(k - 1) = 0 . \tag{8.30}$$

Equation (8.30) is a "difference equation" for $P(k)$, and (again by direct substitution) the reader may verify that it possesses solutions of the form r_1^k and r_2^k, where r_1 and r_2 are the two roots of the quadratic equation

$$r^2 + r\left[-2 + 4\left(\frac{c \, \Delta t}{\Delta x}\right)^2 \sin^2 \frac{\lambda \, \Delta x}{2}\right] + 1 = 0 . \tag{8.31}$$

In fact, the general solution of Equation (8.30) is given by

$$P(k) = B_1 r_1^k + B_2 r_2^k , \tag{8.32}$$

where B_1 and B_2 are constants, since B_1 and B_2 can be chosen so that $P(k)$ has any desired values for $k = 0$ and $k = 1$, and from Equation (8.30) the value of $P(k)$ for $k = 2, 3, \ldots$, is determined uniquely by such a choice. Now if either r_1 or r_2 has an absolute value greater than 1, Equation (8.32) shows that $P(k)$ will grow without limit as k increases. Thus, for stability r_1 and r_2 must have absolute values less than or equal to 1, and the formula for the roots of a quadratic equation shows that this will be the case only if

$$\left|-1 + 2\left(\frac{c \, \Delta t}{\Delta x}\right)^2 \sin^2 \frac{\lambda \, \Delta x}{2}\right| \leq 1$$

or, since λ can have any value,

$$\frac{c \, \Delta t}{\Delta x} \leq 1 . \tag{8.33}$$

This stability condition for the wave equation is known as the *Courant–Friedrichs–Lewy (or CFL) condition.*

The stability constraint (8.33) may be avoided by rephrasing Equation (8.28) in suitable implicit terms. One method is to replace the righthand side by the average of two terms of the same form, one evaluated at time level $k - 1$ and the other at time level $k + 1$. The resulting implicit system is easily solved because of the tridiagonal coefficient matrix.

Many wave-type equations encountered in practice are of more complicated form than Equation (8.24), and the numerical procedure of Equation (8.28) must be appropriately modified. The book by Richtmeyer and Morton (1967) may be consulted. Also, just as in Section 8.2, the number of space dimensions may be greater than 1.

Other methods for obtaining a numerical solution to wave-type equations exist. Some are based on the finite element approach of Section 8.5; we describe here one other, and illustrate it by application to the one-dimensional compressible flow of a gas. If ρ, m, and e denote the density, momentum per unit volume, and energy per unit volume of a gas, all functions of position x and time t, then it is shown in fluid mechanics texts that in the absence of viscosity the motion of the gas is governed by

$$\frac{\partial \rho}{\partial t} = -\frac{\partial m}{\partial x} ,$$

$$\frac{\partial m}{\partial t} = -\frac{\partial}{\partial x}\left[\frac{m^2}{\rho} + p\right] ,$$

$$\frac{\partial e}{\partial t} = -\frac{\partial}{\partial x}\left[(e + p)\frac{m}{\rho}\right] . \tag{8.34}$$

The pressure p is considered to be a known function of e and of ρ from a thermodynamic equation of state. Note that the righthand sides of Equations (8.34) are all in the form $(\partial/\partial x) [\cdots]$; we say Equations (8.34) are then in *conservation form* because of the analogy to conventional conservation laws that link time derivatives to space derivatives. There is some evidence that numerical approximations to equations in conservation form can effectively model rapid transitions, such as in the neighborhood of a shock wave. If V denotes the column vector of the three quantities ρ, m, and e, then Equations (8.34) may be written compactly as

$$\frac{\partial V}{\partial t} = \frac{\partial}{\partial x} F(V) , \tag{8.35}$$

where the vector function $F(V)$ is defined by the righthand sides of Equations (8.34). The *Lax–Wendroff–Richtmeyer method* for Equation (8.35) is now as follows. Denote by V_j^k the numerical approximation to V at the jth space mesh point after k time intervals have passed. Let Δx and Δt be the mesh intervals in space and time. Then each time step is divided into two parts. In the first, intermediate values $V_{j+1/2}^{k+1/2}$ are obtained from

$$V_{j+1/2}^{k+1/2} = \frac{1}{2}(V_j^k + V_{j+1}^k) + \frac{\Delta t}{2\Delta x}[F(V_{j+1}^k) - F(V_j^k)] \qquad (8.36)$$

(the reason for the factor 2 in the term $\Delta t/(2\Delta x)$ is that the time step is $\Delta t/2$). The second part computes the final values for the time step:

$$V_j^{k+1} = V_j^k + \frac{\Delta t}{\Delta x}[F(V_{j+1/2}^{k+1/2}) - F(V_{j-1/2}^{k+1/2})] . \qquad (8.37)$$

This combination of calculations is of second-order accuracy. The method is explicit, and a stability criterion similar to the CFL condition of Equation (8.33) must be satisfied.

8.4 AN EQUATION OF POTENTIAL TYPE

The reader will recall that, in electrostatics, the potential V of a charge Q placed at the origin of (x, y, z) space is given by

$$V = \frac{Q}{\sqrt{x^2 + y^2 + z^2}}$$

in appropriate units. Differentiation shows that

$$\frac{\partial^2 V}{\partial x^2} + \frac{\partial^2 V}{\partial y^2} + \frac{\partial^2 V}{\partial z^2} = 0 ,$$

which is termed *Laplace's equation*. By superposition, the free-space potential arising from any collection of charges satisfies this same equation, and we say that Laplace's equation is an equation of potential type. If the charge distribution permeates space, then the potential satisfies a similar equation,

$$\frac{\partial^2 V}{\partial x^2} + \frac{\partial^2 V}{\partial y^2} + \frac{\partial^2 V}{\partial z^2} = f(x, y, z) , \qquad (8.38)$$

where f is related to the intensity of charge distribution. Equation (8.38) is termed *Poisson's equation*. A two-dimensional version of this equation is applicable to the case in which the charge distribution is independent of the z-coordinate, and this is the example equation we will deal with here.

Let $\phi(x, y)$ satisfy the Poisson equation

$$\frac{\partial^2 \phi}{\partial x^2} + \frac{\partial^2 \phi}{\partial y^2} = f(x, y) \qquad (8.39)$$

in a region R of (x, y) space, on the boundary of which ϕ is prescribed (this kind of boundary condition is called a *Dirichlet condition*). Suppose first that R is a rectangle that can be covered by a square mesh of side h as shown in Figure (8.5). Let $\phi_{i,j}$ denote the numerical approximation to ϕ at the point whose mesh coordinates are (i, j). Then the natural approximation to Equation (8.39) is

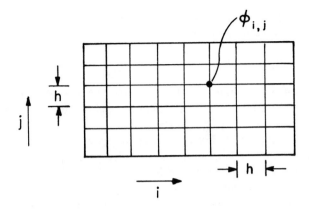

Figure 8.5. Mesh for Poisson equation.

$$\frac{\phi_{i,j+1} - 2\phi_{i,j} + \phi_{i,j-1}}{h^2} + \frac{\phi_{i+1,j} - 2\phi_{i,j} + \phi_{i-1,j}}{h^2} = f_{i,j}, \qquad \textbf{(8.40)}$$

where $f_{i,j}$ denotes the mesh point value of $f(x, y)$. Equation (8.40) is to hold for each interior mesh point (i, j); given boundary values of $\phi_{i,j}$ are to be used as needed.

The number of linear algebraic equations of the form (8.40) is equal to the number of interior mesh points, and frequently this is large enough that an iterative solution of Equation (8.40) is preferable to a direct solution, say by Gaussian elimination. A suitable iterative version of Equation (8.40) is obtained by rewriting it as

$$\phi_{i,j} = \tfrac{1}{4}[(\phi_{i,j+1} + \phi_{i,j-1} + \phi_{i+1,j} + \phi_{i-1,j}) - h^2 f_{i,j}]. \qquad \textbf{(8.41)}$$

Either the Jacobi or the Gauss–Seidel method of Section 2.6 can now be used; in the Jacobi method, "old" values of $\phi_{i,j}$ are used on the righthand side of Equation (8.41), and the lefthand side represents the "new" value of $\phi_{i,j}$ resulting from the iteration process. In the Gauss–Seidel method, which can be expected to be more efficient, "new" values of $\phi_{i,j}$ are used on the righthand side of Equation (8.41) as they become available. Observe that the iterative process is a very natural one; at each point in succession the value of $\phi_{i,j}$ is adjusted to make Equation (8.41) locally exact.

Long before the days of computers, when this kind of iteration was carried out by hand, it was observed that it was often effective to alter groups of ϕ-values as a unit—a process called *block relaxation*. It was also noticed that it was usually useful to overcorrect the value of $\phi_{i,j}$ at each point during an iteration sweep—that is, to compute the required change in $\phi_{i,j}$ at a particular point and to then change $\phi_{i,j}$ by more than this amount. Block relaxation can be achieved on a computer by making occasional use of a coarser mesh, interpolating back and forth between the coarse mesh and the fine mesh. As far as over relaxation is concerned, let us rewrite Equation (8.41) in the form

$$\phi_{i,j} = \phi_{i,j} + \frac{A}{4} [\phi_{i,j+1} + \phi_{i,j-1} + \phi_{i+1,j} + \phi_{i-1,j} - 4\phi_{i,j} - h^2 f_{i,j}] , \quad \textbf{(8.42)}$$

where the constant A represents the relaxation factor. Again, all of the $\phi_{i,j}$ values on the righthand side of Equation (8.42) would be "old" values if Jacobi relaxation were being used, but would include some "new" values in the case of Gauss–Seidel relaxation. With $A = 1$, Equation (8.41) is recovered; a choice of A greater than 1 corresponds to over relaxation, and a choice of A less than 1 corresponds to under relaxation. It turns out that with the "optimal" choice for A, convergence of the iterative process is rather dramatically enhanced; for example, the number of iteration sweeps (using Gauss–Seidel relaxation) required to decrease an initial error by a factor of 100 is now of the order of N (the number of mesh points along a side) rather than N^2, as would be the case for $R = 1$. The optimal value of A can be calculated for a rectangle (it is just less than 2), but for a more general region it must be obtained by experimentation (and may in fact have different values in different parts of the region). This method is called the *successive optimal over-relaxation method*, or *SOR method*.

If the boundary is curved, then approximations of the kind described in Section 8.2 may be made. If the boundary conditions involve derivatives of ϕ in the normal direction (i.e., perpendicular to the boundary) as well as ϕ itself, then the form of Equation (8.42) must be appropriately modified adjacent to a boundary point. For a bit more nomenclature, if the normal derivative of ϕ on the boudary is specified everywhere, the problem is said to be of *Neumann type*. In a Neumann problem it is clear that any constant can be added to a solution ϕ without affecting any of the equations satisfied by ϕ, so the solution to a Neumann problem is only unique within an additive constant.

8.5 THE FINITE ELEMENT IDEA

In the finite element method the region of interest is divided into subregions of simple geometrical shape, and the unknown function is assumed to be of partic-ular form inside each subregion, this form being characterized by parameters. The overall problem then becomes one of determining these parameters. Although finite elements can be used in many contexts, the illustrative application here will be to a problem of potential equation type.

Let $\phi(x, y)$ satisfy the Poisson equation

$$\frac{\partial^2 \phi}{\partial x^2} + \frac{\partial^2 \phi}{\partial y^2} = f(x, y) \qquad \textbf{(8.43)}$$

in the region R shown in Figure 8.6, where $f(x, y)$ is prescribed, and where values of ϕ on the boundary Γ of R are given. Divide R into triangular subregions, at

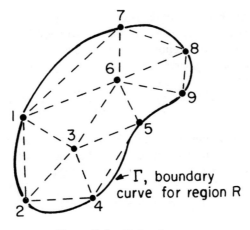

Figure 8.6. Finite elements.

least approximately, and number the nodes arbitrarily. Denote the values of ϕ at node j by ϕ_j. Values of ϕ_j at boundary nodes are assigned in accordance with the prescribed boundary data, so in this simple example only ϕ_3 and ϕ_6 are unknown. In any one triangle, ϕ is assumed to have linear form, so that, for example, the value of ϕ inside the triangle with nodes 1, 2, 3, is assumed to be given by

$$\phi = ax + by + c \, ,$$

where the coefficients a, b, c may be expressed as linear combinations of the nodal values ϕ_1, ϕ_2, and ϕ_3 (thus,

$$a = \frac{\phi_1(y_2 - y_3) + \phi_2(y_3 - y_1) + \phi_3(y_1 - y_2)}{2A} \, ,$$

where A is the area of the triangle and x_j, y_j are the coordinates of the jth node, etc.).

At this point it is useful to rephrase the problem of Equation (8.43) in the following equivalent form: "Find a function ϕ which takes on prescribed boundary values on Γ and is such that

$$I = \int_R \left[\left(\frac{\partial \phi}{\partial x} \right)^2 + \left(\frac{\partial \phi}{\partial y} \right)^2 + 2f\phi \right] dA \tag{8.44}$$

is as small as possible,* considering all such functions ϕ."

---

*We say "as small as possible" for ease of description. Actually, as we consider small changes in ϕ that do not violate the boundary conditions, we only require that the corresponding first-order change in ϕ should vanish. Although this does often correspond to a minimum for I, it might represent a maximum or a point of inflection—much as in conventional maxima and minima.

It is easy to sketch a proof of the equivalence of the two problems formulated in Equations (8.43) and 8.44), if we for the moment cover R with a square mesh of side h in the manner of Section 8.4. Then temporarily using the notation of Section 8.4 and denoting the area of a mesh square by ΔA,

$$I \cong \sum \left\{ \left[\frac{\phi_{i+1,j} - \phi_{i,j}}{h} \right]^2 + \left[\frac{\phi_{i,j+1} - \phi_{i,j}}{h} \right]^2 + 2f_{i,j}\phi_{i,j} \right\} \Delta A ,$$

and the condition for a minimum is that $\partial I/\partial \phi_{ij} = 0$ for each interior mesh point value $\phi_{i,j}$. Considering the various terms in the sum in which a particular $\phi_{i,j}$ appears, this leads to

$$-\frac{2}{h^2} (\phi_{i+1,j} - \phi_{i,j}) + \frac{2}{h^2} (\phi_{i,j} - \phi_{i-1,j})$$

$$-\frac{2}{h^2} (\phi_{i,j+1} - \phi_{i,j}) + \frac{2}{h^2} (\phi_{i,j} - \phi_{i,j-1}) + 2f_{i,j} = 0 ,$$

which coincides with Equation (8.40), thus showing equivalance.

Since α is assumed to have the form $ax + by + c$ inside each triangle (where, of course, a, b, and c will have different values in different triangles), $\partial \phi/\partial x$ and $\partial \phi/\partial y$ are constant inside each triangle, being determined as linear combinations of the nodal values ϕ_j for that triangle. The integral over each triangle of $(\partial \phi/\partial x)^2 + (\partial \phi/\partial y)^2$ becomes a quadratic form in the nodal values ϕ_j. The contribution of $2f\phi$ to the integral is also easily evaluated in terms of the nodal values of f and ϕ_j (e.g., assume f also to be a linear function of x and y in each triangle, determined by the nodal values). Approximating the integral I by a sum of contributions over all triangles, we obtain an overall quadratic form in the ϕ_j values, which is to be minimized (or, more precisely, made stationary). The derivative of the quadratic form with respect to each interior ϕ_j value is set equal to zero, and this gives a set of linear algebraic equations to be solved for the interior ϕ_j values. (In the example of Fig. 8.6, only two equations, involving the two unknowns ϕ_3 and ϕ_6, would result.)

Geometrical figures other than triangles can be used, and the form of ϕ within any such figure can be given a more complicated form. One advantage of the finite element method is that, by using triangles, say, the boundary of a particular region may be "fitted" better than by the use of a rectangular mesh.

ANNOTATED BIBLIOGRAPHY

E. B. Becker, G. F. Carey, and J. T. Oden, 1981, *Finite Elements, An Introduction*, vol. 1, Prentice-Hall, Englewood Cliffs, New Jersey, 258p.

This short text gives a clear treatment of a variety of practical topics in the field of finite elements. Particular attention is directed towards computational aspects.

G. Birkhoff and R. E. Lynch, 1984, *Numerical Solution of Elliptic Problems*, Society of Industrial and Applied Mathematics, Philadelphia, 319p.

This book supplements Richtmeyer and Morton (1967) in that it concerns potential-type rather than evolution-type equations. It considers integral equation and variational methods as well as finite difference methods. An extensive bibliography is included.

R. D. Richtmeyer and K. W. Morton, 1967, *Difference Methods for Initial Value Problems*, 2nd ed., Wiley-Interscience, New York, 405p.

This book is divided into two parts. Part I is preoccupied with Banach spaces and a number of other good things, but fortunately Part II, *Applications*, can be read without knowledge of Part I. Chapter 8 is particularly worthwhile; it includes a number of computer experiments on "time evolution" equations of diffusion type. Later chapters deal with neutron transport and wave-type equations.

G. D. Smith, 1978, *Numerical Solution of Partial Differential Equations: Finite Difference Methods*, 2nd ed., Oxford University Press, Oxford, England, 179p.

This is a fairly elementary treatment of numerical methods for classical types of partial differential equations, enhanced by the inclusion of a number of worked examples.

PROBLEMS

8.1 Write a program that implements Equation (8.12) for the unit interval, and verify the stability condition (8.14) experimentally. Use $\phi = 0$ at the ends and choose any initial condition [e.g., $\phi(x, 0) = x(x - l)$].

8.2 Use the sine function method (as was done for Equation (8.12) to show that Equation (8.17) is unstable and that Equation (8.18) is stable for any choices of Δx, Δt.

8.3 Write a computer program that solves Equation (8.18), using the tridiagonal matrix method of Problem 2.13. Let $\phi(0, t)$ and $\phi(l, t)$ be specified arbitrarily, and let $\phi(x, 0)$ also be arbitrary. Run some trial problems, including some for which an exact solution is available. Monitor accuracy by the use of a halved mesh.

8.4 A 1-meter thick wall, made of a material for which $\alpha^2 = .1$ cm^2/sec, is initially at 20°C. At time $t = 0$ one side of the wall is suddenly subjected to a steady temperature of 300°C; the other side is maintained at 20°C. Use a numerical method (e.g., the result of Prob. 8.3) to determine the time history of the temperature profile in the wall and of the heat transfer through the wall. Check your solution for large t against the easily found steady-state solution corresponding to $t = \infty$.

8.5 Repeat Problem 8.4 for the case in which the colder wall is insulated, so that $\partial \phi / \partial x = 0$ there.

8.6 A simple method (often used) for finding boundary values of ϕ at a typical point A in Figure 8.3 is to just use the value of ϕ at that point on the original curved boundary that is closest to A. Can you do better than this, using, for example, the interpolation ideas of Chapter 5? Many approaches are possible; list some reasonable ideas.

8.7 An alternative to the Peachman–Rachford–Douglas scheme of Section 8.2 is to replace the righthand side of Equation (8.21) by the finite difference equivalent of $2(\partial^2\phi/\partial x^2)$ in the first of each pair of steps, and by $2(\partial^2\phi/\partial y^2)$ in the second; an implicit formulation is used each time. Write a computer program that implements this *Godunov–Bagrinovskii method* for a rectangle, on the border of which ϕ is a prescribed function of time and where initial values of ϕ are given inside the rectangle. Devise a test procedure. How could this method be extended to three dimensions?

8.8 Write out the finite difference approximation to the two-dimensional wave equation

$$\frac{\partial\phi}{\partial t} = c^2\left(\frac{\partial^2\phi}{\partial x^2} + \frac{\partial^2\phi}{\partial y^2}\right),$$

where $c = $ constant. Derive the CFL condition that now corresponds to Equation (8.33)

8.9 Solve Laplace's equation (with Dirichlet boundary conditions) in a square region by the Jacobi, Gauss–Seidel, and SOR methods. In the latter, determine the relaxation parameter by experimentation, using boundary conditions that correspond to a simple known solution (like $\phi = x^2 - y^2$). Compare and discuss.

8.10 Show that Equation (8.24) can be written in conservation form. [*Hint*: Define $u = \partial\phi/\partial x$, $v = \partial\phi/\partial t$, etc.] Write a computer program that uses the Lax–Wendroff–Richtmeyer method [cf. Eqs.(8.36) and (8.37)] applied to this form to solve Equation (8.24) for the case in which $\phi(0, t) = \phi(l, t) = 0$ and in which $\phi(x, 0)$ is specified. Compare the results with the direct method of Equation (8.28).

8.11 An alternative formulation of a numerical approximation to Equation (8.4) is to discretize only in space, so as to obtain the coupled set of ordinary differential equations

$$\frac{d\phi_j}{dt} = \frac{\alpha^2}{(\Delta x)^2}[\phi_{j+1} - 2\phi_j + \phi_{j-1}],$$

which may be solved by, say, a Runge–Kutta method. Explore this idea for some trial cases. Also investigate the possibility of a similar approach to the wave equation. How about higher-dimensional problems?

8.12 Suppose that an infinite string is given an initial deformation extending over a finite x-interval, and the string is then released from rest. Solve the problem numerically, and observe that the initial deformation pattern splits into two equal halves, moving in opposite directions with velocity c. (Can you derive this result analytically?) If the string is finite, the moving deformation pattern gets reflected at the fixed endpoint; investigate this reflection numerically.

8.13 In solving Equations (8.41) by iteration, it is also possible to carry out the iterations line by line, alternating sweeps between the horizontal and vertical directions. Investigate and discuss.

8.14 Laplace's equation for a circular region in terms of polar coordinates r and θ reads

$$\frac{\partial^2 \phi}{\partial r^2} + \frac{1}{r}\frac{\partial \phi}{\partial r} + \frac{1}{r^2}\frac{\partial^2 \phi}{\partial \theta^2} = 0 \ .$$

Write a program that solves the Dirichlet problem for a circle and apply it to some test cases. How do you handle the central point?

8.15 Carry out analytically the details of the finite element method for Laplace's equation (with Dirichlet boundary conditions) in the rectangle of Figure 8.7, using the triangular subdivision shown, and obtain the resulting finite difference equations that the nodal values of ϕ must satisfy. Compare these with the conventional finite difference equations for this problem.

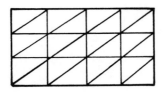

Figure 8.7. Geometry of Problem 8.15.

TAYLOR'S THEOREM

Taylor was a student of Newton, so his theorem dates from the early days of calculus. It gives a relation between values of a function $f(x)$ at two different points lying in an interval I within which the first $n + 1$ derivatives of $f(x)$ exist and are continuous.

Let a be some fixed point in I, and let x be some other point in I. Then Taylor's theorem (in modern form) states that

$$f(x) = f(a) + (x - a)f'(a) + \frac{1}{2!}(x - a)^2 f''(a) + \cdots$$

$$+ \frac{1}{n!}(x - a)^n f^{(n)}(a) + R_n(x),$$

where the remainder term $R_n(x)$ is given by

$$R_n(x) = \frac{1}{(n + 1)!}(x - a)^{n+1} f^{(n+1)}(\xi).$$

Here ξ (which depends on x) lies in the interval (a, x). Derivatives of f are indicated by primes; thus f' means df/dx, f'' means $d^2 f/dx^2$, and so on.

To derive this result, write $f(x) - f(a) = \int_a^x f'(\beta) \, d\beta$ and integrate by parts; the integrand has the form $u \, dv$, where $u = f'(\beta)$, and we take $v = \beta - x$. (Remember that a and x are fixed; the variable of integration is β.) We obtain

$$f(x) - f(a) = [f'(\beta) \cdot (\beta - x)]_a^x - \int_a^x f''(\beta)(\beta - x) \, d\beta$$

$$= (x - a)f'(a) + R_1(x).$$

The expression for $R_1(x)$ may again be integrated by parts, using $u = f''(\beta)$, $v = (\beta - x)^2/2!$; continuation of this process yields

$$f(x) - f(a) = (x - a)f'(a) + \frac{1}{2!}(x - a)^2 f''(a) + \cdots$$

$$+ \frac{1}{n!}(x - a)^n f^{(n)}(a) + R_n(x) ,$$

where

$$R_n(x) = \frac{1}{n!} \int_a^x (x - \beta)^n f^{(n+1)}(\beta) \, d\beta .$$

But $(x - \beta)^n$ does not change sign as β varies between a and x, so $R_n(x)$ must equal the integral of the term $(x - \beta)^n/n!$ multiplied by some mean value of $f^{(n+1)}(\beta)$. Since this latter function is continuous, its mean value must be attained at some point ξ in (a, x); hence,

$$R_n(x) = f^{(n+1)}(\xi) \left\{ \frac{1}{n!} \int_a^x (x - \beta)^n \, d\beta \right\} ,$$

and this now gives the stated result.

In numerical analysis the remainder term $R_n(x)$ is frequently discarded. This approximation is valid provided that $|x - a|$ is sufficiently small, since we require $f^{(n+1)}$ to be continuous in the (closed) interval I and, therefore, bounded. Then since R_n involves a higher power of $x - a$ than the other terms, it may be made as relatively small as desired by choosing $|x - a|$ to be sufficiently small.

A simple example is given by $f(x) = \cos x$, $a = 0$, $n = 3$. Taylor's theorem gives

$$\cos x = 1 + 0 \cdot x - \frac{1}{2!} x^2 + 0 \cdot x^3 + \frac{1}{4!} x^4 \cos \xi .$$

Since $|\cos \xi| < 1$, it is clear that, for small values of x, the approximation $\cos x \cong 1 - \frac{1}{2} x^2$ is a very good one.

For this example we can continue the expansion indefinitely. Since it is clear that, for any value of x, $R_n(x) \to 0$ as $n \to \infty$, it follows that the series of alternating terms

$$\cos x = 1 - \frac{x^2}{2!} + \frac{x^4}{4!} - \frac{x^6}{6!} + \cdots$$

converges for any value of x. The result is termed a *Taylor series*. Other examples, derived just as easily, are

$$\sin x = x - \frac{x^3}{3!} + \frac{x^5}{5!} - \frac{x^7}{7!} + \cdots ,$$

$$\exp x = 1 + x + \frac{x^2}{2!} + \frac{x^3}{3!} + \frac{x^4}{4!} + \cdots ,$$

$$\sinh x = x + \frac{x^3}{3!} + \frac{x^5}{5!} + \frac{x^7}{7!} + \cdots,$$

$$\cosh x = 1 + \frac{x^2}{2!} + \frac{x^4}{4!} + \frac{x^6}{6!} + \cdots,$$

$$\ln(1 + x) = x - \frac{x^2}{2} + \frac{x^3}{3} - \frac{x^4}{4} + \frac{x^5}{5} - \cdots.$$

The expansion for the logarithm converges for $|x| < 1$; the others converge for all values of x.

DETERMINANTS

With epsilon and delta symbols, determinant theory can be presented quite compactly; we begin by defining these symbols. They are then used to obtain the properties of determinants of order 3, and it is pointed out in a final section that very little change is required to carry everything over to the case of a determinant of order n.

B.1 PERMUTATION SYMBOL AND SUMMATION CONVENTION

Consider some permutation of the natural integers from 1 to n. Suppose this permutation is brought into the natural order $1, 2, 3, \ldots, n$ by a sequence of pair interchanges. Then it turns out that the required number of such interchanges will be either always odd or always even. For example, take $n = 4$ and let the particular punctuation be 3, 1, 4, 2. Then, interchanging one pair of numbers at a time, each of the following sequences will reestablish the natural order:

$$(3, 1, 4, 2) \rightarrow (1, 3, 4, 2) \rightarrow (1, 2, 4, 3) \rightarrow (1, 2, 3, 4),$$

$$(3, 1, 4, 2) \rightarrow (4, 1, 3, 2) \rightarrow (1, 4, 3, 2) \rightarrow (1, 3, 4, 2)$$

$$\rightarrow (1, 3, 2, 4) \rightarrow (1, 2, 3, 4).$$

The first sequence required three interchanges; the second, five. The number of each interchange is odd, and any other sequence of interchanges would also be an odd number.

To prove the general result, denote the first integer in the original permutation by i_1, the second by i_2, and so on. Then the original permutation is $(i_1, i_2, i_3, \cdots, i_n)$. Associate with any such permutation the product

$$P(i_1, i_2, \cdots, i_n) = [(i_1 - i_2)(i_1 - i_3) \cdots (i_1 - i_n)]$$
$$\cdot \, [(i_2 - i_3)(i_2 - i_4) \cdots (i_2 - i_n)] \cdots [(i_{n-1} - i_n)] \, .$$

Then the interchange of any pair is easily seen to reverse the sign of P (try an example to see how the argument goes); since the value of P associated with the natural order is of fixed sign, it follows at once that the total number of interchanges will be always even or always odd. The original permutation is termed an *even* permutation of the natural order or an *odd* permutation, respectively.

Let each of $i_1, i_2, \cdots, i_n$ be some integer in the range $(1, 2, \cdots, n)$. Then the *permutation symbol* $\epsilon_{i_1 i_2 \cdots i_n}$ is defined to vanish, unless all of the integers from 1 to n are actually included in the subscripts; in that case it equals $+1$ or -1 according to whether $(i_1, i_2, \cdots, i_n)$ is an even or an odd permutation of the natural order.

Thus, if $n = 4$, we have

$$\epsilon_{2314} = +1 \, , \qquad \epsilon_{3112} = 0 \, ,$$
$$\epsilon_{1234} = +1 \, , \qquad \epsilon_{1324} = -1 \, ,$$
$$\epsilon_{4114} = 0 \, , \qquad \epsilon_{2134} = -1 \, ,$$

and so on.

Next consider the expression

$$S_{ip} = \sum_{j=1}^{3} \sum_{k=1}^{3} \epsilon_{ijk} \epsilon_{pjk} \, ,$$

where i and p may be any of the values 1, 2, 3. Suppose first that i and p are both 1. Then

$$S_{ip} = \epsilon_{123} \epsilon_{123} + \epsilon_{132} \epsilon_{132} \, ,$$

where the vanishing terms (such as $\epsilon_{122} \epsilon_{122}$) have been omitted. Thus, in this case $S_{11} = 2$. Next, let $i = 1$, $p = 2$. There are now no nonvanishing terms in the sum; if the one factor does not vanish, the other does. Thus $S_{12} = 0$. In fact, we easily find, in general, that $S_{ij} = 0$ unless i and j are the same integer, in which case $S_{ij} = 2$. A convenient symbol used to represent this state of affairs is the quantity δ_{ij} (sometimes called the Kronecker delta), which is defined to equal unity if i and j are the same integer, and zero otherwise. Thus

$$\sum_{j=1}^{3} \sum_{k=1}^{3} \epsilon_{ijk} \epsilon_{pjk} = 2\delta_{ip} \, . \tag{B.1}$$

More generally, and by the same kind of argument, it follows that

$$\sum_{j=1}^{n} \sum_{k=1}^{n} \cdots \sum_{s=1}^{n} \epsilon_{ijk\cdots s}\epsilon_{pjk\cdots s} = (n-1)! \cdot \delta_{ip} . \tag{B.2}$$

In determinant theory one often encounters sums over indices of this character. To avoid having to continually write summation signs, it is common practice to adopt the *summation convention*, in which a summation from 1 to n is understood to be associated with any *repeated* index. Thus, Equation (B.1) will be written

$$\epsilon_{ijk}\epsilon_{pjk} = 2\delta_{ip} ,$$

where summation from 1 to 3 is understood to be applied to each of the repeated indices j and k. Similarly, Equation (B.2) will in the future be written

$$\epsilon_{ijk\cdots s}\epsilon_{pjk\cdots s} = (n-1)! \, \delta_{ip} .$$

In this book the summation convention (often attributed to Einstein, who encountered many summations in his tensor formulation of relativity theory) will be used in this appendix only.

B.2 DETERMINANT OF ORDER 3

Let A denote a 3×3 square array of elements a_{ij}:

$$A = \begin{bmatrix} a_{11} & a_{12} & a_{13} \\ a_{21} & a_{22} & a_{23} \\ a_{31} & a_{32} & a_{33} \end{bmatrix} .$$

Here the element a_{ij} in the ith row and jth column may be any real or complex number. Then using the permutation symbol ϵ_{ijk} of the preceding section, we define the determinant of A, written a, or det (A), or $|A|$, by

$$a = \epsilon_{ijk}a_{1i}a_{2j}a_{3k} . \tag{B.3}$$

Since each of the indices i, j, and k is repeated, there are three understood summations from 1 to 3 in this expression, so the righthand side of Equation B.3 is really the sum of 27 terms. Each possible combination of integers i, j, k from 1 to 3 occurs exactly once in this sum; of course, the factor ϵ_{ijk} will vanish if any two of the i, j, k indices are the same.

The reader should check that when the sum is written out and the ϵ_{ijk} quantities are evaluated, then

$$a = a_{11}a_{22}a_{33} + a_{13}a_{21}a_{32} + a_{12}a_{23}a_{31}$$

$$-a_{13}a_{22}a_{31} - a_{12}a_{21}a_{33} - a_{11}a_{23}a_{32} .$$

This corresponds to adding the products indicated by the solid lines and subtracting those indicated by the dotted lines in the array

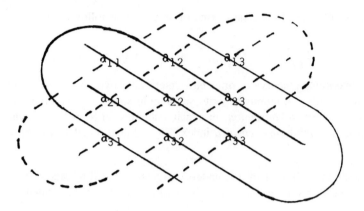

In Equation (B.3) observe that each nonvanishing product involves one element from each row and one from each column; with the rows in natural order, the sign of each such product is $+1$ or -1, depending on whether the column order is an even or odd permutation of the natural order.

Equation (B.3) may be generalized to become

$$\epsilon_{pqr} a = \epsilon_{ijk} a_{pi} a_{qj} a_{rk} , \tag{B.4}$$

where p, q, r may be any integers in the range (1, 2, 3). To verify Equation (B.4), observe first that if $p = 1$, $q = 2$, and $r = 3$, then Equation (B.3) is recovered. If $p = 2$, $q = 1$, $r = 3$, then Equation (B.4) reads

$$-a = \epsilon_{ijk} a_{2i} a_{1j} a_{3k} .$$

If we replace the summation index i by j, and j by i, this equation becomes

$$-a = \epsilon_{jik} a_{2j} a_{1i} a_{3k} = -\epsilon_{ijk} a_{2j} a_{1i} a_{3k} ,$$

which agrees with Equation (B.3). Finally, if $p = q = 1$ and $r = 3$, both sides of Equation B.4 are zero, which is easily seen by writing out the summation over i and j. We complete the argument by considering the other possible index values, and Equation (B.4) is verified.

Define next

$$\alpha = \epsilon_{ijk} a_{i1} a_{j2} a_{k3} , \tag{B.5}$$

which differs from a as defined in Equation (B.3) in the order of the indices of the elements a_{ij}. Recall again that there is an implied summation over each of i, j, k. We will prove that $\alpha = a$. Note first that, just as in the generalization of Equation (B.3) to Equation (B.4), Equation (B.5) leads to

$$\epsilon_{pqr} \alpha = \epsilon_{ijk} a_{ip} a_{jq} a_{kr} . \tag{B.6}$$

To prove that $a = \alpha$, simply multiply each side of Equations (B.4) and (B.6) by ϵ_{pqr} and sum over each of p, q, and r. This gives

$$6a = \epsilon_{pqr} \epsilon_{ijk} a_{pi} a_{qj} a_{rk} \quad \text{and} \quad 6\alpha = \epsilon_{pqr} \epsilon_{ijk} a_{ip} a_{jq} a_{kr} ,$$

respectively. In the righthand sides of these two equations, we have summations over all pairs of indices, and the two results are clearly the same. This completes the proof.

An interesting property of a determinant is that if we add a multiple of one row to any other row, the value of the determinant does not change. To exemplify the proof, let us add C times the second row to the third row. Then we want to show that

$$\epsilon_{ijk} a_{1i} a_{2j} (a_{3k} + C a_{2k}) = a$$

or, equivalently, that

$$\epsilon_{ijk} a_{1i} a_{2j} a_{2k} = 0 .$$

But this is certainly true, since in the sum over j and k each nonvanishing term will appear twice, with opposite signs. The general proof proceeds similarly. The same statement holds if the word *row* is replaced by the word *column*.

To multiply two determinants, let

$$b = \epsilon_{ijk} b_{i1} b_{j2} b_{k3} ,$$

so that

$$
\begin{aligned}
ab &= (a \epsilon_{ijk}) b_{i1} b_{j2} b_{k3} \\
&= (\epsilon_{pqr} a_{pi} a_{qj} a_{rk}) b_{i1} b_{j2} b_{k3} \\
&= \epsilon_{pqr} (a_{pi} b_{i1}) (a_{qj} b_{j2}) (a_{rk} b_{k3})
\end{aligned}
$$

This is the determinant of a new array whose elements are formed by the matrix multiplication rule of Chapter 2.

B.3 COFACTORS AND CRAMER'S RULE FOR ORDER 3 CASE

Multiply each side of Equation (B.4) by ϵ_{sqr} and sum over the repeated indices q and r. Using Equation (B.1), we get

$$\delta_{sp} a = a_{pi} (\tfrac{1}{2} \epsilon_{sqr} \epsilon_{ijk} a_{qj} a_{rk}) .$$

The quantity in parenthesis is called the *cofactor* of a_{si}; we denote it by A_{si}. Thus

$$A_{si} = \tfrac{1}{2} \epsilon_{sqr} \epsilon_{ijk} a_{qj} a_{rk} , \tag{B.7}$$

and we have

$$a_{pi} A_{si} = \delta_{sp} a . \tag{B.8}$$

For example,

$$A_{12} = \tfrac{1}{2} \, \epsilon_{1qr} \epsilon_{2jk} a_{qj} a_{rk}$$

$$= \tfrac{1}{2} \left[\epsilon_{123} \epsilon_{2jk} a_{2j} a_{3k} + \epsilon_{132} \epsilon_{2jk} a_{3j} a_{2k} \right] \, .$$

But the second term here equals the first (rename the j's as k's and vice versa, and observe that $\epsilon_{2kj} = -\epsilon_{2jk}$), so that

$$A_{12} = \epsilon_{2jk} a_{2j} a_{3k} \, ,$$

which is the value of the determinant of an array found from the original (a_{ij}) array by replacing a_{12} by unity and all other elements in the first row by zeros (since this modified determinant would equal $\epsilon_{ijk} \delta_{i2} a_{2j} a_{3k}$, and summation over i gives the above result). Generalizing this argument, we observe that A_{ij} equals the value of the determinant obtained by replacing all a_{ij} in the ith row by zeros, except for a_{ij} itself, which is replaced by unity. (*Note:* Instead of modifying the ith row in this manner, we could modify the jth column analogously.)

Equation (B.8) states that the sum of the products of the elements in any row by the cofactors of any row is zero if these rows are different, and is a if they are the same.

Starting with Equation (B.6) (with $\alpha = a$), instead of Equation (B.4), we obtain a relation analogous to Equation (B.8), but which may be interpreted as a sum over the elements in a column rather than a row:

$$a_{ip} A_{is} = \delta_{ps} a \, . \tag{B.9}$$

Finally, we derive Cramer's rule. A set of three linear algebraic equations in the three unknowns (x_1, x_2, x_3) can be written in the form

$$a_{ij} x_j = b_i$$

for $i = 1, 2, 3$, where a_{ij} and b_i are given numbers and the summation convention applies to the repeated index j. Multiply both sides by A_{ip} (summing over the repeated index i) to get

$$A_{ip} a_{ij} x_j = A_{ip} b_i$$

or, from Equation (B.9),

$$a \delta_{pj} x_j = A_{ip} b_i \, ;$$

that is,

$$a x_p = A_{ip} b_i \, . \tag{B.10}$$

This is *Cramer's rule*. Recalling the previous interpretation of A_{ip} (in column form), we see that the righthand side is equivalent to the calculation of a determinant involving the a_{ij} with the pth column replaced by the b_i elements.

B.4 DETERMINANTS OF ANY ORDER

The derivations of the various results concerning third-order determinants carry over directly to higher-order systems; it is only necessary to replace Equation (B.1) by Equation (B.2). Thus, with n elements,

$$a = \epsilon_{i_1 i_2 \cdots i_n} a_{1 i_1} a_{2 i_2} \cdots a_{n i_n}$$

$$= \epsilon_{i_1 i_2 \cdots i_n} a_{i_1 1} a_{i_2 2} \cdots a_{i_n n} .$$

Also

$$\epsilon_{i_1 i_2 \cdots i_n} a = \epsilon_{j_1 j_2 \cdots j_n} a_{i_1 j_1} a_{i_2 j_2} \cdots a_{i_n j_n} ,$$

$$\epsilon_{i_1 i_2 \cdots i_n} a = \epsilon_{j_1 j_2 \cdots j_n} a_{j_1 i_1} a_{j_2 i_2} \cdots a_{j_n i_n} ,$$

$$A_{pq} = \frac{1}{(n-1)!} \epsilon_{p i_2 i_3 \cdots i_n} \epsilon_{q j_2 j_3 \cdots j_n} a_{i_2 j_2} a_{i_3 j_3} \cdots a_{j_n i_n} ,$$

and

$$a_{ip} A_{jp} = a \delta_{ij} , \qquad a_{pi} A_{pj} = a \delta_{ij} .$$

With reference to Chapter 2, observe that the inverse matrix to the array (a_{ij}), if it exists, is the transposed matrix of cofactors, each cofactor divided by a.

COMPLEX NUMBERS

C.1 DEFINITIONS

A complex number z will be taken to have the general form $x + iy$, where x and y are real. We say x is the *real part* of z and y is the *imaginary part* of z; we write $x = \text{Re}(z)$ and $y = \text{Im}(z)$. The number z may be depicted geometrically as a point in the (x, y) plane; see Figure C.1. In this figure the x-axis is termed the real axis, and the y-axis is termed the imaginary axis. The *modulus* or *absolute value* of z, written $|z|$, is the nonnegative real number defined by $|z| = (x^2 + y^2)^{1/2}$. The *argument* of z, written $\arg(z)$, is the angle measured from the positive x-axis to the radius vector drawn from the origin to the point z; by convention, the counterclockwise direction is taken as positive. If $\theta = \arg(z)$, then it is seen from Figure C.1 that $z = |z|(\cos \theta + i \sin \theta)$. The value of θ associated with any chosen point z is determined within an arbitrary multiple of 2π radians.

Two complex numbers are equal if their real and imaginary parts are respectively equal. The sum or difference of two complex numbers $z_1 = x_1 + y_1$ and $z_2 = x_2 + y_2$ is defined by

$$z_1 \pm z_2 = (x_1 \pm x_2) + i(y_1 \pm y_2) .$$

The geometrically obvious *triangle inequality*, $|z_1 + z_2| \leq |z_1| + |z_2|$, is often used. The product of z_1 and z_2 is defined by

$$z_1 z_2 = (x_1 + iy_1)(x_2 + iy_2) = (x_1 x_2 - y_1 y_2) + i(x_1 y_2 + x_2 y_1) .$$

The usual distributive and associative laws hold, so that, for example, $z_1(z_2 + z_3) = z_1 z_2 + z_1 z_3$, and $(z_1 z_2) z_3 = z_1(z_2 z_3)$. A useful result that follows from the above

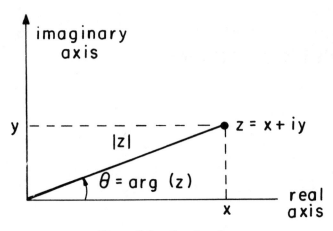

Figure C.1. Complex plane.

definitions is that if $z_3 = z_1 z_2$, then $|z_3| = |z_1| \cdot |z_2|$, and $\arg(z_3) = \arg(z_1) + \arg(z_2)$, within a multiple of 2π.

The complex conjugate z^* of a number $z = x + iy$ is defined by $z^* = x - iy$. The quotient z_1/z_2, if z_2 is nonzero, is defined to be that number z_3 satisfying $z_1 = z_2 z_3$; if this equation is interpreted in terms of real and imaginary parts, it is seen that z_3 will be determined uniquely. To find z_3 in practice, it is often convenient to multiply the numerator and denominator of the fraction z_1/z_2 by z_2^*. For example,

$$\frac{3 - i}{1 + 2i} = \frac{(3 - i)(1 - 2i)}{(1 + 2i)(1 - 2i)} = \frac{1 - 7i}{5} = \frac{1}{5} - \frac{7}{5} i .$$

Let $A_1, A_2, \ldots$ be a sequence of complex numbers. We say that this sequence converges to the limit L (again a complex number) if $|A_n - L| \to 0$ as $n \to \infty$. We write $A_n \to L$ as $n \to \infty$. Next, consider the infinite series of complex numbers $u_1 + u_2 + u_3 + \cdots$. Define s_n to be the sum of the first n terms. Then we say the series converges to a sum L if $s_n \to L$ as $n \to \infty$.

An infinite series of particular interest is the complex equivalent of the exponential function of Appendix A. Define

$$E(z) = 1 + z + \frac{1}{2!} z^2 + \frac{1}{3!} z^3 + \cdots . \tag{C.1}$$

Assign to z some chosen complex value. Then the triangle inequality shows that the absolute value of the sum of any finite collection of terms in this series is not greater than the sum of the individual absolute values. However, $E(|z|)$ is just the familiar exponential function of the real variable $|z|$, and since this series converges, then (because of the dominance just noted) $E(z)$ itself must converge. Thus $E(z)$ converges for any complex value assigned to z.

Direct substitution together with a little algebra shows that if z_1 and z_2 are any complex numbers then

$$E(z_1 + z_2) = E(z_1) \cdot E(z_2) . \tag{C.2}$$

As a special case, let z_1 equal x and z_2 equal iy, where x and y are real. Then using $E(x) = e^x$ (see Appendix A), we obtain

$$E(x + iy) = E(x) \cdot (E(iy))$$

$$= e^x \left\{ \left(1 - \frac{1}{2!} y^2 + \frac{1}{4!} y^4 + \cdots \right) + i \left(y - \frac{y^3}{3!} + \frac{y^5}{5!} - \cdots \right) \right\}$$

$$= e^x [\cos y + i \sin y] , \tag{C.3}$$

where the Taylor series for $\cos y$ and $\sin y$ of Appendix A have been used in the last step.

Equation (C.3) implies that, given any complex number ξ, we can find another complex number $z = x + iy$ such that $E(z) = \xi$. In fact, we need only choose x such that $e^x = |\xi|$ and $y = \arg(\xi)$. Of course, any integral multiple of 2π could be added to or subtracted from y, and we would still have $E(z) = \xi$. The number z obtained in this way is termed the logarithm of ξ; we write $z = \ln \xi$.

We are now in a position to define complex powers of complex numbers. Let a and b be complex. Then, by definition,

$$a^b = E(b \ln a) . \tag{C.4}$$

This definition is certainly valid if a and b are real numbers, since E is then the usual exponential function, and

$$e^{b \ln a} = (e^{\ln a})^b = a^b .$$

Consequently, what we are doing in Equation (C.4) is simply carrying over the real-variable definition to the complex case, using a form in which the expressions are defined for complex quantities.

The fact that $\ln a$ is only determined within a multiple of 2π means that complex powers are, in general, multiple-valued. For example, $\ln (-1) = i\pi \pm 2\pi in$, where n is an arbitrary integer, so that

$$(-1)^{1/3} = E\left[\frac{1}{3}(i\pi \pm 2\pi in) \right] = E\left[i\frac{\pi}{3} \pm \frac{2}{3} i\pi n \right] .$$

Using Equation (C.3) and considering all possible values of n, we find that there are three possible values for $(-1)^{1/3}$: $\frac{1}{2}(1 + i\sqrt{3})$, $\frac{1}{2}(1 - i\sqrt{3})$, and -1. Similarly, the reader may check that $(1 + i)^3$ has only one value, and that i^i has the possible values $e^{-\pi/2 \pm 2\pi n}$.

A final remark here is that, for obvious reasons, it is conventional to denote $E(z)$ by e^z.

C.2 DERIVATIVES OF COMPLEX FUNCTIONS

Suppose that $w = f(z)$ is a complex-valued function of a complex variable z, in the sense that a value of $f(z)$ is associated with each value of z, for values of z lying in some region in the complex plane of Figure C.1.

Consider a fixed point z, and let $z + \Delta z$ be a neighboring point. Then if the ratio

$$\frac{f(z + \Delta z) - f(z)}{\Delta z}$$

approaches a limit L as $\Delta z \to 0$, we say that $f(z)$ has a derivative at the point z, and that the value of the derivative is L (L is often denoted by df/dz or $f'(z)$).

Note that Δz, the complex displacement from z of the neighboring point, can be in any direction from z; it is required that the same limit be obtained as $\Delta z \to 0$ from any direction. Because the definition of a derivative has the same algebraic form as in the real-variable case, the usual rules for differentiation of a product or a quotient hold. We will not pursue the general topic of complex differentiation (and integration) further here, but we will consider some examples:

$$\frac{d}{dz}(z^3) = \lim_{\Delta z \to 0} \frac{(z + \Delta z)^3 - z^3}{\Delta z}$$

$$= \lim_{\Delta z \to 0} \frac{1}{\Delta z}[z^3 + 3z^2\Delta z + 3z(\Delta z)^2 + (\Delta z)^3 - z^3]$$

$$= 3z^2 \, ,$$

which has the same form as in the real-variable case; more generally, it is clear than $d(z^n)/dz = nz^{n-1}$.

For z complex, adapt the Taylor series expansions for $\sin x$ and $\cos x$ to define

$$\sin z = z - \frac{1}{3!}z^3 + \frac{1}{5!}z^5 \cdots \, ,$$

$$\cos z = 1 - \frac{1}{2!}z^2 + \frac{1}{4!}z^4 \cdots \, ,$$

each of which is easily seen to be convergent.

Term-by-term differentiation is valid here (for the same reasons as in the real-variable case), and we find at once that

$$\frac{d}{dz}(\sin z) = \cos z, \qquad \frac{d}{dz}(\cos z) = -\sin z \, .$$

For the exponential function $E(z)$ we similarly find

$$\frac{d}{dz}[E(z)] = E(z) \, .$$

C.3 ZEROS OF POLYNOMIALS

If z is a nonzero complex number, then $\arg(z)$ can have an infinite number of values differing from one another by multiples of 2π. Suppose we choose one of these values. Now make z traverse some path in the z-plane such that $\arg(z)$ varies continuously along this path (i.e., we do not permit $\arg(z)$ to jump at any point by a multiple of 2π). We do not let the path pass through the origin since $\arg(z)$ is there undefined. Then if the path eventually returns to its starting point, it is clear that $\arg(z)$ will eventually return either to its initial value or to that value incremented by some multiple of 2π (otherwise, z could not return to its original value). Moreover, if the path of the moving point has not encircled the origin, then $\arg(z)$ will return exactly to its original value; otherwise $\arg(z)$ will increment by 2π for each counterclockwise encirclement of the origin and will decrement by 2π for each clockwise encirclement of the origin. The reader should visualize the situation by starting with Figure C.1 and watching what happens to θ as z moves.

Next, let $w = Az^2$, where A is a complex constant. A similar visualization, with $\arg(w) = \arg(A) + \arg(z^2)$, where $\arg(z^2) = 2\arg(z)$, shows that if we require $\arg(z)$ to vary continuously along the path, then, upon return to the starting point, we find that $\arg(w)$ has incremented by 4π for each counterclockwise encirclement of the origin and decremented by 4π for each clockwise encirclement. Finally, if $w = Az^n$, where n is some positive integer, the above statement continues to hold if we replace 4π by $2n\pi$.

Now let

$$w(z) = a_0 + a_1 z + a_2 z^2 + \cdots + a_n z^n$$

be a polynomial of degree n, where the a_j are constants. We want to prove that there is at least one value of z for which $w(z) = 0$; this result is often called the *fundamental theorem of algebra*. Before proceeding, we remark that this result will lead at once to the conclusion that a polynomial of degree n has exactly n zeros, for if it has one zero at $z = z_1$, say, then by the usual division process of algebra we can write $w(z) = (z - z_1) P_{n-1}(z)$, where $P_{n-1}(z)$ is a polynomial of degree $n - 1$. But the theorem now requires P_{n-1} to have at least one zero, say at $z = z_2$, and the obvious recursion process now leads to the result that $w(z)$ has exactly n zeros; hence,

$$w(z) = (z - z_1)(z - z_2) \cdots (z - z_n) .$$

In general, the z_i will be complex numbers, which need not all be distinct.

To prove the fundamental theorem itself, suppose the contrary. But if $w(z)$ has no zeros, then $\arg(w)$ is well defined at every point (within the usual multiple of 2π). Suppose we draw a very large circle centered on the origin, so large that the term $a_n z^n$ makes the dominant contribution to the value of w. As we traverse this curve in the counterclockwise direction (here and in the future requiring $\arg(w)$ to vary continuously), it is then clear that $\arg(w)$ will increment by $2\pi n$. Now shrink the circle very slightly; continuity requires that $\arg(w)$ still increment by $2\pi n$ (it certainly must increment by some multiple of 2π, since w has to come back to its original value, and any other multiple than n of 2π is ruled out by continuity). Proceeding in this way, we eventually find that arg(w) increments by $2\pi n$ as we follow a circle of minute radius around the origin, and this provides the desired contradiction since arg(z) must itself be continuous—it can change only slightly for a slight change in z. Thus, the assumption that $w(z)$ vanish nowhere must be false, and $w(z)$ must have at least one zero. As pointed out previously, it then follows that a polynomial of degree n must in fact have exactly n zeros.

D

PARTIAL DERIVATIVES

D.1 DEFINITION

Let $w = f(x, y)$ be a function of the two variables x and y. By the phrase *partial derivative of w with respect to x* (denoted by $\partial w/\partial x$), we mean exactly the same thing as the ordinary derivative of w with respect to x, except that the variable y is to be held fixed during the differentiation. Thus, if $w = x^2 - y^3 + \sin(x - y^2)$, then

$$\frac{\partial w}{\partial x} = 2x + \cos(x - y^2) .$$

Similarly,

$$\frac{\partial w}{\partial y} = -3y^2 - 2y \cos(x - y^2) ,$$

where the differentiation was carried out with respect to y, the x-variable being held fixed.

Higher derivatives are formed in the same way. Using the same example, we obtain

$$\frac{\partial^2 w}{\partial x^2} = \frac{\partial}{\partial x}\left(\frac{\partial w}{\partial x}\right) = 2 - \sin(x - y^2) ,$$

$$\frac{\partial^2 w}{\partial y^2} = \frac{\partial}{\partial y}\left(\frac{\partial w}{\partial y}\right) = -6y - 2\cos(x - y^2) - 4y^2 \sin(x - y^2) .$$

Also,

$$\frac{\partial}{\partial y}\left(\frac{\partial w}{\partial x}\right) = 2y \sin (x - y^2) ,$$

$$\frac{\partial}{\partial x}\left(\frac{\partial w}{\partial y}\right) = 2y \sin (x - y^2) .$$

It is no accident that the last two expressions are identical; it is in fact true that (if the various partial derivatives are continuous) *the order of partial differentiation is immaterial*. To prove this result, let (x, y) be a chosen point and consider the function

$$G = f(x + h, y + k) - f(x + h, y) - f(x, y + k) + f(x, y) ,$$

where h and k represent small alterations in the arguments. Defining $\phi(x, y) = f(x, y + k) - f(x, y)$, we can write G as

$$G = \phi(x + h, y) - \phi(x, y) = \phi_x(x + \alpha h, y) \cdot h ,$$

where ϕ_x denotes $\partial\phi/\partial x$ and $0 < \alpha < 1$; here we have used the mean value theorem from elementary calculus. Thus, using the definition of ϕ, we get

$$G = [f_x(x + \alpha h, y + k) - f_x(x + \alpha h, y)] \cdot h ,$$

and by another application of the mean value theorem we obtain

$$G = f_{xy}(x + \alpha h, y + \beta k) \cdot hk ,$$

where $0 < \beta < 1$. However, we could also have written $\psi(x, y) = f(x + h, y) - f(x, y)$, so that $G = \psi(x, y + k) - \psi(x, y)$; in that case we would have been led to

$$G = f_{yx}(x + \alpha_1 h, y + \beta_1 k) \cdot kh ,$$

with $0 < \alpha_1 < 1, 0 < \beta_1 < 1$. Equating the two expressions for G and letting h and k approach 0, we obtain $f_{xy} = f_{yx}$, and this completes the proof.

In the sequel we need the *chain rule* of differentiation. Suppose $w = f(x, y)$, where x and y are functions of a variable t—say $x = \phi(t)$, $y = \psi(t)$. Then w is implicitly a function of t, and to find dw/dt we consider

$$\frac{1}{\Delta t} [f(x + \Delta x, y + \Delta y) - f(x, y)] ,$$

where $x + \Delta x = \phi(t + \Delta t)$, $y + \Delta y = \psi(t + \Delta t)$. Writing this expression as

$$\frac{1}{\Delta t} [\{f(x + \Delta x, y + \Delta y) - f(x, y + \Delta y)\} + \{f(x, y + \Delta y) - f(x, y)\}]$$

$$= \frac{1}{\Delta t} [f_x(x + \alpha \Delta x, y + \Delta y) \cdot \Delta x + f_y(x, y + \beta \Delta y) \cdot \Delta y] ,$$

where $0 < \alpha < 1, 0 < \beta < 1$, and the mean value theorem has again been used, it is clear that we will obtain, in the limit,

$$\frac{dw}{dt} = f_x \frac{dx}{dt} + f_y \frac{dy}{dt} .$$

This is the chain rule.

Similar results hold for partial derivatives of functions involving more than two variables; the derivations are analogous. For example, if $w = xyz^2 - x^2y$, then $w_x = yz^2 - 2xy$, $w_{xx} = -2y$, $w_{yz} = w_{zy} = 2xz$, and so forth. Also, if x, y, z are functions of t, then

$$\frac{dw}{dt} = (yz^2 - 2xy) \frac{dx}{dt} + (xz^2 - x^2) \frac{dy}{dt} + (2xyz) \frac{dz}{dt} .$$

D.2 TAYLOR'S THEOREM IN TWO OR MORE VARIABLES

Taylor's Theorem of Appendix A may be extended almost immediately to the several-variable case. Consider first $f(x, y)$, evaluated in the neighborhood of a point (a, b). Define $x - a = h, y - b = k$, and (for fixed choices of x, y, h, and k)

$$\phi(t) = f(a + th, b + tk) .$$

We will eventually be interested in the choice $t = 1$, but for the moment think of ϕ as a function of the single variable t.

Then from Taylor's Theorem we have

$$\phi(t) = \phi(0) + t\phi'(0) + R_1(t) ,$$

where

$$R_1(t) = \frac{1}{2!} t^2 \phi''(\xi) ,$$

where ξ is some number in $(0, t)$. Evaluating ϕ, ϕ', and ϕ'', we obtain

$$f(a + th, b + th) = f(a, b) + t\{f_x h + f_y b\}$$
$$+ \tfrac{1}{2} t^2 \{f_{xx} h^2 + 2f_{xy} hk + f_{yy} k^2\} ,$$

where the chain rule of differentiation has been used. Here f_x and f_y are evaluated at (a, b), and f_{xx}, f_{xy}, f_{yy} are evaluated at $(a + \xi h, b + \xi k)$. If h and k are small enough that R_1 is negligible, we find (with $t = 1$) that

$$f(x, y) \cong f(a, b) + f_x \cdot (x - a) + f_y(y - b) .$$

If the next term is included, we find in the same way that

$$f(x, y) \cong f(a, b) + f_x \cdot (x - h) + f_y(y - b)$$

$$+ \frac{1}{2!} [f_{xx}(x - a)^2 + 2f_{xy}(x - a)(y - b) + f_{yy}(y - b)^2],$$

where all partial derivatives are evaluated at (a, b).

The inclusion of one more term makes the general situation clear:

$$f(x, y) \cong f(a, b) + f_x \cdot (x - a) + f_y \cdot (y - b)$$

$$+ \frac{1}{2!} [f_{xx} \cdot (x - a)^2 + 2f_{xy} \cdot (x - a)(y - b) + f_{yy} \cdot (y - b)^2]$$

$$+ \frac{1}{3!} [f_{xxx} \cdot (x - a)^3 + 3f_{xxy} \cdot (x - a)^2 (y - b)$$

$$+ 3f_{xyy}(x - a)(y - b)^2 + f_{yyy} \cdot (y - b)^3].$$

If f is a function of more than two variables, the corresponding formulas may be obtained in the same way. We give the single example

$$f(x, y, z) \cong f(a, b, c) + f_x \cdot (x - a) + f_y \cdot (y - b) + f_z \cdot (z - c)$$

$$+ \frac{1}{2!} [f_{xx} \cdot (x - a)^2 + 2f_{xy} \cdot (x - a)(y - b) + f_{yy} \cdot (y - b)^2$$

$$+ 2f_{xz} \cdot (x - a)(z - c) + 2f_{yz} \cdot (y - b)(z - c) + f_{zz}(z - c)^2].$$

Index

About the Author

CARL E. PEARSON is a professor of applied mathematics at the University of Washington, with a joint appointment in aeronautics and astronautics. He has previously taught at Harvard and at the Technical University of Denmark. He has also held research or managerial positions with Arthur D. Little, Sperry, and Boeing. Dr. Pearson's publications include *Theoretical Elasticity* (Harvard University Press), *Functions of a Complex Variable* (McGraw-Hill), and *Partial Differential Equations* (Academic Press). He is editor of the *Handbook of Applied Mathematics* (Van Nostrand Reinhold).